T0275809

Statistical inference for spatial processes

STATISTICAL INFERENCE FOR SPATIAL PROCESSES

B.D. RIPLEY

Professor of Statistics,
University of Strathclyde

An essay awarded the Adams Prize of the University of Cambridge

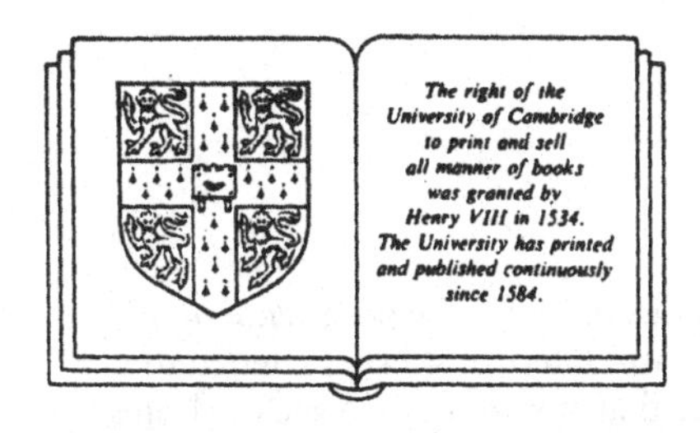

CAMBRIDGE UNIVERSITY PRESS

Cambridge

New York Port Chester Melbourne Sydney

CAMBRIDGE UNIVERSITY PRESS
Cambridge, New York, Melbourne, Madrid, Cape Town, Singapore,
São Paulo, Delhi, Dubai, Tokyo, Mexico City

Cambridge University Press
The Edinburgh Building, Cambridge CB2 8RU, UK

Published in the United States of America by
Cambridge University Press, New York

www.cambridge.org
Information on this title: www.cambridge.org/9780521424202

First published 1988
First paperback edition 1991

A catalogue record for this publication is available from the British Library

Library of Congress Cataloguing in Publication Data

Ripley, Brian D., 1952–
Statistical inference for spatial processes/B.D. Ripley.
p. cm.
Bibliography: p.
Includes index.
ISBN 0 521 35234 7 – ISBN 0 521 42420 8 (pb)
1. Spatial analysis (Statistics) I. Title.
QA278.2.R57 1988 87-35489

ISBN 978-0-521-35234-5 Hardback
ISBN 978-0-521-42420-2 Paperback

Contents

Preface

The statistical study of spatial patterns and processes has during the last few years provided a series of challenging problems for theories of statistical inference. Those challenges are the subject of this essay. As befits an essay, the results presented here are not in definitive form; indeed, many of the contributions raise as many questions as they answer. The essay is intended both for specialists in spatial statistics, who will discover much that has been achieved since the author's book (*Spatial Statistics*, Wiley, 1981), and for theoretical statisticians with an eye for problems arising in statistical practice.

This essay arose from the Adams Prize competition of the University of Cambridge, whose subject for 1985/6 was 'Spatial and Geometrical Aspects of Probability and Statistics'. (It differs only slightly from the version which was awarded that prize.) The introductory chapter answers the question 'what's so special about spatial statistics?' The next three chapters elaborate on this by providing examples of new difficulties with likelihood inference in spatial Gaussian processes, the dominance of edge effects for the estimation of interaction in point processes. We show by example how Monte Carlo methods can make likelihood methods feasible in problems traditionally thought intractable.

The last two chapters deal with digital images. Here the problems are principally ones of scale dealing with up to a quarter of a million data points. Chapter 5 takes a very general Bayesian viewpoint and shows the importance of spatial models to encapsulate prior information about images. As well as original contributions it contains an overview of statistical work on image segmentation. Chapter 6 is concerned rather with summarizing and understanding images, the beginnings of an 'exploratory data analysis' for images. This is important both in the evaluation of the algorithms presented in chapter 5 and in many applications in biology, geology and materials science.

Personal computers with good graphical displays are making most of the techniques described here available at low cost. I would encourage readers to experiment for themselves.

ACKNOWLEDGEMENTS

The work reported in chapters 5 and 6 was supported by the UK Science and Engineering Research Council. Mr Jeremy Warnes computed the figures for chapter 2. Chapter 3 is based on collaborative work with Drs Ohser and Stoyan of the Bergakademie, Freiberg, DDR, and I thank them for bringing some of the problems to my attention. Section 5.3 reports joint work with Rafael Molina Soriano on problems raised by his colleagues at Instituto de Astrofisica de Andalucia, Granada, Spain. Finally, Mrs Lynne Westwood prepared the typescript with great speed and efficiency.

B. D. Ripley
Glasgow, June 1987

1

Introduction

This essay aims to bring out some of the distinctive features and special problems of statistical inference on spatial processes. Realistic spatial stochastic processes are so far removed from the classical domain of statistical theory (sequences of independent, identically distributed observations) that they can provide a rather severe test of classical methods. Although much of the literature has been very negative about the problem, a few methods have emerged in this field which have spread to many other complex statistical problems. There is a sense in which spatial problems are currently the test bed for ideas in inference on complex stochastic systems.

Our definition of 'spatial process' is wide. It certainly includes all the areas of the author's monograph (Ripley, 1981), as well as more recent problems in image processing and analysis. Digital images are recorded as a set of observations (black/white, greylevel, colour...) on a square or hexagonal lattice. As such, they differ only in scale from other spatial phenomena which are sampled on a regular grid. Now the difference in scale *is* important, but it has become clear that it is fruitful to regard imaging problems from the viewpoint of spatial statistics, and this has been done quite extensively within the last five years.

Much of our consideration depends only on geometrical aspects of spatial patterns and processes. Two of the fundamental difficulties of spatial processes are the lack of any causal ordering of the observations, and the simple observation that

$$\frac{\text{number of `neighbours' at distance} \leqslant t}{t} \rightarrow \infty$$

as $t \rightarrow \infty$. Both difficulties divide the study of spatial processes from that of time series. In the 1950s there was an optimistic view that time series methods could be extended simply to spatial processes (e.g. Whittle, 1954). This was accepted for a long time, and the error in the assumption (although rather simple to demonstrate) was forcefully rebutted only by Guyon's (1982) careful treatment of the spatial problem. From the perspective of 1986 the chasm appears to be between

classical and time series methods

on the one hand, and

spatial and complex system methods

on the other.

The difficulties of inference for spatial processes can conveniently be grouped under a small number of headings.

(a) Edges

Consider $\{1, \ldots, n\}^d \subset \mathbb{Z}^d$. This contains $N = n^d$ points, of which $[n^d - (n-2)^d]$ are 'outside' points, in that they have fewer than $2d$ neighbours. The number of outside points $N_0 \sim 2dn^{d-1} = 2dN^{1-(1/d)}$ and so

$$\frac{N_0}{N} \sim 2dN^{-1/d}$$

Suppose we had independent observations at each point of the lattice, and computed some measure of the similarity of the observation at each point to those of its $2d$ neighbours. We would expect the statistic so computed to converge in distribution at rate $N^{-1/2}$, by the central limit theorem. The problem of spatial processes is that for $d \geqslant 2$ the error term due to the outside points is of at least the same order as the statistical fluctuation term. Far from being asymptotically negligible, edge effects often dominate the asymptotic distribution, and almost always dominate problems of typical size.

(b) Which way to infinity?

In classical problems and in time series it is totally clear how to embed the problem uniquely in a series of problems so as to perform asymptotic calculations. This is not at all the case in spatial problems. Consider an $m \times n$ portion of $\mathbb{Z}^2$. Clearly either m or n should tend to infinity, but very little can be concluded about the ratio m/n. The reader could be forgiven for assuming that this might be immaterial, and in some problems it is. However, there are problems (Ripley, 1982, p. 252) in which the asymptotic distribution has a mean depending on the asymptotic value of m/n!

These problems have a further twist for spatial point processes. Suppose we observe n points irregularly distributed within a study region E. One way to embed this in a sequence of problems is to let $n \to \infty$ and keep E fixed. This will have the effect of increasing the interaction between the points, and the edge effects will be O(1). Perhaps a more natural approach is to suppose that the pattern extends throughout space but was only observed within E. Then an asymptotic sequence of problems will be to take a sequence of 'windows' E increasing to $\mathbb{R}^d$. This keeps the interaction rate fixed but diminishes the edge effects. The ways in which these different asymptotic regimes give different limit results are studied in chapters 3 and 4.

(c) Long-range dependence

In time series analysis short-range dependence is the norm and special models are needed to demonstrate long-range dependence. In spatial problems long-range dependence appears inevitable. This is best illustrated by a few examples.

(i) Moving average contouring methods. Suppose we wish to interpolate a continuous surface $Z(\ )$, observed at n points $\mathbf{x}_1, \ldots, \mathbf{x}_n \in \mathbb{R}^d$. A very simple way to do so is to use a weighted average

$$\hat{Z}(\mathbf{x}) = \sum_1^n \lambda_i(\mathbf{x}) Z(\mathbf{x}_i), \qquad \Sigma \lambda_i(\mathbf{x}) = 1$$

with

$$\lambda_i(\mathbf{x}) \propto w(d(\mathbf{x}, \mathbf{x}_i))$$

Clearly the fitted surface will interpolate if $w(r) \to \infty$ as $r \to 0$. We would also like $\hat{Z}(\ )$ to be differentiable, not least so as to be able to contour it without difficulty. This needs $r/w(r) \to 0$ as $r \to 0$ (Ripley, 1981, p. 36). On the other hand, let us consider large r. We would expect $O(r^{d-1})\Delta r$ points at distances r to $r + \Delta r$ away from $\mathbf{x}$, and these contribute $O(r^{d-1} w(r))\Delta r$ to the total of the weights. Thus unless

$$\int_1^\infty r^{d-1} w(r) dr < \infty$$

then $\hat{Z}()$ will not be a local average but will depend entirely on the size of the study region.

This algorithm has been used quite widely in contouring packages for twenty years. One common choice of $w(r)$ was r^{-2}, which gives a smooth (C^2) interpolated surface, but with long-range dependence in $\mathbb{R}^2$.

(ii) Boundary conditions. The important distinction between a process defined only on the study region E and one defined throughout $\mathbb{Z}^d$ or $\mathbb{R}^d$ but observed in E has often been overlooked. The difference is one of boundary conditions. For example, consider a simple conditional autoregression (Ripley, 1981, p. 88), a Gaussian process with

$$\mathrm{E}(Z_i \mid \text{rest}) = \alpha \sum_{j \text{ nhbr of } i} Z_j$$

For the process defined only for a finite part of the lattice the outside points will have fewer neighbours, but the effect is identical to setting $Z_i \equiv 0$ for $i \notin E$. Again, consider a point process of centres of discs randomly packed within E. If the process extends throughout space the discs will be forced away from the edges of E by discs whose centres lie outside E and so are unobserved. Both the proportion of edge sites *and* the geometry of the problem mean that the differences in the boundary conditions have an effect throughout E and not just near the edges.

(iii) Lattice processes. Consider again the conditional autoregression. This has direct dependence only between neighbouring sites. On $\{\ldots, -1, 0, 1, \ldots\}$ the connection between points n time units apart is through all intermediate points, and so the correlation decays as ρ^n. In $\mathbb{Z}^d$, $d > 1$, this is no longer the case. There are many paths between points n steps apart on the lattice, and the number of paths increases with n. This can balance the decay along each individual path, so that correlations decay quite slowly with n. It also ensures that a process which is defined on a grid can appear quite isotropic at medium and large scales. Besag (1981) gives some detailed calculations to support these assertions.

This remark is connected with the phenomenon of *phase transition* in statistical physics. Consider the simple model on the lattice $\mathbb{Z}^d$ which takes values ± 1 at each point. Let

$$\mathrm{P}(Z_{ij} = +1 \mid \text{all other values}) = \beta \sum_{\text{nhbrs}} Z_{rs}$$

where there are $2d$ neighbours. Then on $\mathbb{Z}$ the process has characteristics continuous in β. For $d \geqslant 2$ there is a *critical point* β_c at which many characteristics have a discontinuity. In particular, for $\beta > \beta_c$ the correlation between values at sites distance n apart does not decrease to zero with n, but converges to some positive constant, and realizations almost surely have infinite patches of $+1$ and of -1. For further details of this process see Pickard (1987).

(d) Geometry of likelihoods

There are essentially two problems with likelihood inference for spatial processes. The best known is computational. There are problems in which it is impossible to write the likelihood in a simple closed form, and others in which the form is simple but the combinatorial terms involved are prohibitive. We give two examples from spatial point processes.

(i) Cluster processes. Suppose we have m parent points distributed independently and uniformly within E. Around each parent there is a random number n (with distribution $f(n)$) of daughter points, each of which is independently distributed about the parent with a uniformly distributed orientation and radial distance p.d.f. $P_R(\;)$. Then the p.d.f. of the n observed daughter points is

$$p(\mathbf{x}_1,\ldots,\mathbf{x}_n)=\sum f(n_i)\prod_{j=1}^{n_i} P_R(\|\mathbf{x}_{ij}-\mathbf{y}_i\|)$$

where the sum is over the assignment of the observed daughter points to the parents $\mathbf{y}_1,\ldots,\mathbf{y}_m$. Since m and the location of the parents are usually unknown, this expression will need to be averaged over $\mathbf{y}_i$ uniform in E and over m. The combinatorial sum is prohibitive even for moderate n: there are m^n terms!

(ii) Gibbsian point processes are a way to model interactions between points, for instance to allow spacing out of birds' nests. One class of models gives a p.d.f. to n points as

$$p(\mathbf{x}_1,\ldots,\mathbf{x}_n)=\frac{1}{Z}\prod_{i<j} h(d(\mathbf{x}_i,\mathbf{x}_j))$$

where

$$Z=\int_{E^n}\prod_{i<j} h(d(\mathbf{x}_i,\mathbf{x}_j))\,\Pi\, d\mathbf{x}_i$$

To perform likelihood inference we need to know Z as a function of the parameters in h. This is a problem in which some progress has been made both in approximating Z and in estimating Z (Monte Carlo inference), discussed further in chapter 4.

There is, however, a more fundamental problem. Until recently it has been widely assumed that likelihood methods are in some sense near-optimal in spatial problems. (About the only published dissenting voice is Ripley, 1984b.) There is no theoretical basis for this belief. Classical theorems on strong consistency and best asymptotic normality apply to sequences of independent identically distributed (i.i.d.) random variables. The work of Mann and Wald extended these to time series, but in a context in which there is a sequence of i.i.d. innovations. There are efficiency results for spatial processes (e.g. Mardia and Marshall, 1984) but these depend on embedding the problem in an asymptotic sequence of almost independent copies, so it is no surprise that the classical results are obtained. As we saw in (b) 'which way to infinity?', this particular asymptotic formulation need not provide useful guidance for even moderately sized problems.

The author has been suspicious of these results for some time, but only the very recent examples presented in chapter 2 demonstrated the scale of the problem. It appears likely that similar difficulties in fact occur in many other spatial contexts.

(e) Stationarity

Some assumption of stationarity plays a crucial role in virtually all of spatial statistics. Despite the fact that this is almost axiomatic it has often been argued against by users. Statistical inference is impossible without *some* stationarity assumption. Most spatial problems have only one data set and replication has to be attained from stationarity. It is this that the users misunderstand; they inconsistently argue against stationarity and simultaneously compute average characteristics of their data sets. If stationarity is false then these average characteristics have no meaning!

There is still a worthwhile debate to be had on what stationarity assumption should be made. Perhaps the easiest way to explain the difficulties is to draw parallels with time series analysis. There departures from stationarity are usually trends in mean level. (Trends in variance are studied occasionally.) These trends are usually of a simple form such as a smooth increase and/or seasonal variation, and are typically removed by differencing, perhaps after an instantaneous transformation of the data (say to log scale). Some differences are:

(i) Spatial differencing is not comprehensive. Künsch (1987) shows that the class of differenced autoregressions in $\mathbb{R}^d$, $d \geqslant 2$, is not a natural generalization of autoregressions, and some important limits of spatial autoregressions are omitted. This difficulty, like several others, stems from the lack of factorizability of polynomials over $\mathbb{C}^d$ for $d \geqslant 2$.

(ii) The class of possible departures is very large. Suppose we look at a spatial point pattern and decide it is not stationary. We could have noticed

trend in intensity from top to bottom
trend in any other direction
'banding' from periodic variation in intensity in any direction
'patchiness' on one of many scales

at least. In fact, we will usually notice something even in simulations from stationary processes! The problem is essentially that of multiple comparisons. We are implicitly applying tens or even hundreds of significance tests to the pattern we see, and reporting the most significant.

(iii) Stationarity can be under translations and/or rotations. As well as the complications of having two separate ideas, *homogeneity* (stationarity under translations) and *isotropy* (stationarity under rotations about a fixed point, with homogeneity, about any point), there is a structural problem. It is rather convenient that with stationarity under translations the group is isomorphic to the underlying space ($\mathbb{Z}^d$ or $\mathbb{R}^d$). The trivial observation that this is untrue for the group of rigid motions in $\mathbb{R}^d$ has nontrivial consequences.

One attempt to overcome these problems is the *intrinsic hypothesis of* Matheron (1973), which applies to a real-valued surface in $\mathbb{R}^d$ (and hence specializes to lattice processes). A generalized increment of order k is $\Sigma\lambda_i Z(\mathbf{x}_i)$, the weights λ_i satisfying

$$\Sigma\lambda_i x_{i1}^{\alpha_1}\dots x_{id}^{\alpha_d}=0 \qquad \forall\alpha_i\geqslant 0,\quad \Sigma\alpha_i\leqslant k$$

where $\mathbf{x}_i=(x_{i1},\dots,x_{id})$. The hypothesis is that all generalized increments processes of order k are stationary. The increment of order k filters out polynomials of order k, so a process can have a polynomial trend and still satisfy the intrinsic hypothesis. This avoids the problem of polynomial trends by differencing them away.

Another approach, currently favoured by the author, is to model trends as another layer of stochastic variation. We then have the stationary model

$$Z(\mathbf{x})=Z_m(\mathbf{x})+Z_l(\mathbf{x})$$

where Z_l has short-range dependence but Z_m has long-range dependence and so its realizations resemble trends. This doubly stochastic approach has the advantage that we do not need to specify what trends can occur, and unlike polynomial trend surfaces, it extrapolates safely.

(f) Discretization

Like many other applications of stochastic processes, the non-point-process part of spatial statistics is concerned with sampled or aggregated data. So is time series, but there are two important differences. First, even when the underlying continuous process is isotropic, the sampled version will not be. The option of modelling the continuous phenomenon and basing inference on the sampled or aggregated data is usually computationally prohibitive. Easy models on a rectangular lattice mostly do not give even approximately isotropic realizations. This problem is mainly recognized, then ignored.

The second problem is that whereas time series are usually sampled regularly, spatial phenomena are not. (Even the varying lengths of months and quarters in time series are usually ignored.)

The above catalogue of problems may give a rather bleak impression, but this would be incorrect. It is intended rather to show why spatial problems are different and challenging. The rest of this essay shows how some of these challenges have been addressed.

2

Likelihood analysis for spatial Gaussian processes

The framework used in this chapter was developed by the author (Ripley, 1981, chapter 4) to provide a formal statistical framework for the ideas of Matheron and his school of 'geostatistiques'. The ideas provide a generalization to spatial problems of the Wiener–Kolmogorov theory of prediction in time series, and provide a flexible framework for smoothing and interpolation of spatial surfaces.

We suppose that a surface $Z(\mathbf{x})$ is defined for $\mathbf{x} \in X \subset \mathbb{R}^d$. This could be topographic height over a geographical region or porosity in a (three-dimensional) oil reserve. The surface is assumed to be smooth, at least continuous and preferably differentiable. The most tractable model is a spatial Gaussian process. This is defined by the mean function

$$m(\mathbf{x}) = \mathrm{E}Z(\mathbf{x})$$

and covariance function

$$c(\mathbf{x}, \mathbf{y}) = \mathrm{cov}[Z(\mathbf{x}), Z(\mathbf{y})]$$

plus joint normality of the finite-dimensional distributions. Let us parameterize the mean function by a spatial regression model as

$$m(\mathbf{x}) = f(\mathbf{x})^{\mathrm{T}}\beta$$

Now suppose the surface is observed at $\mathbf{x}_1, \ldots, \mathbf{x}_n$ and we wish to predict the surface elsewhere. The minimum mean square error unbiased predictor $\hat{Z}(\mathbf{x})$ is given by

$$\hat{Z}(\mathbf{x}) = y^{\mathrm{T}}k(\mathbf{x}) + f(\mathbf{x})^{\mathrm{T}}\hat{\beta}$$

where

$$K = [c(\mathbf{x}_i, \mathbf{x}_j)] \qquad k(\mathbf{x}) = [c(\mathbf{x}, \mathbf{x}_i)]$$

$$F = \begin{bmatrix} f(\mathbf{x}_1)^{\mathrm{T}} \\ \vdots \\ f(\mathbf{x}_n)^{\mathrm{T}} \end{bmatrix} \qquad \mathbf{Z} = \begin{bmatrix} Z(\mathbf{x}_1) \\ \vdots \\ Z(\mathbf{x}_n) \end{bmatrix}$$

LL^T is the Cholesky decomposition of K, so L is lower triangular.

$L^{-1}\mathbf{Z} = L^{-1}\mathbf{F}\hat{\beta}$ gives $\hat{\beta}$.

$L(L^T y) = [Z(\mathbf{x}_i) - f(\mathbf{x}_i)^T \hat{\beta}]$ gives y.

This is a computationally stable form of 'universal kriging'. Further,

$$\operatorname{var}[Z(\mathbf{x}) - \hat{Z}(\mathbf{x})] = c(\mathbf{x}, \mathbf{x}) - \|\mathbf{e}\|^2 + \|\mathbf{g}\|^2 \tag{1}$$

$$L\mathbf{e} = k(\mathbf{x})$$

$$R^T\mathbf{g} = f(\mathbf{x}) - (L^{-1}F)^T\mathbf{e}$$

R is the orthogonal reduction of $L^{-1}F$

This is a model-based approach which is both its strength and its weakness. The freedom to choose $c(\ ,\ )$ gives great flexibility to the technique. Conversely, c is never known and must be estimated from the data. This is done in *ad hoc* ways by the geostatistiques school. Since examples in Ripley (1981, chapter 4) and Warnes (1986) show that rather small changes in c can give rise to large changes in the fitted surface $\hat{Z}(\ )$, it is clear that the prediction variance given by (1) can very seriously underestimate the true prediction uncertainty. This is analogous to using a regression equation for prediction whilst ignoring the variability of the regression coefficients.

Spatial Gaussian processes have also been used as models for the error term of a spatial regression, particularly in geography and for agricultural field trials (Ripley, 1981, chapter 5; Cook and Pocock, 1983; Besag and Kempton, 1986). The model is again

$$Z(\mathbf{x}) = m(\mathbf{x}) + \varepsilon$$

$$m(\mathbf{x}) = f(\mathbf{x})^T\beta \tag{2}$$

but no prediction is involved, so $\mathbf{x}$ is restricted to n sites $\mathbf{x}_1, \ldots, \mathbf{x}_n$. An example might be to explain health variables on small geographical units by environmental factors. The errors $(\varepsilon_1, \ldots, \varepsilon_n)$ are assumed to be spatial autocorrelated from other environmental factors not explicitly included in the regression (2). Thus

$$\boldsymbol{\varepsilon} \sim \text{MVN}(\mathbf{0}, K)$$

where MVN denotes a multivariate normal distribution. Again $c(\ ,\ )$ and hence K is unknown and has to be fitted from the data. A parametric form $c(\mathbf{x}, \mathbf{y}; \theta)$ is assumed, and in this field maximum likelihood has been proposed for the estimator of (β, θ). The log likelihood is

$$\mathrm{L}(\beta, \theta; \mathbf{Z}) = \text{const} - \tfrac{1}{2}[\ln|K_\theta| + (\mathbf{Z} - F\beta)^T K_\theta^{-1}(\mathbf{Z} - F\beta)] \tag{3}$$

We ignore the constant from now on. The MLE of β minimizes the quadratic form, and so is the generalized least squares estimate $\hat{\beta}$ given by

$$L^{-1}\mathbf{Z} = L^{-1}F\hat{\beta}$$

as before. Thus the profile likelihood for θ is

$$L_p(\theta; \mathbf{Z}) = -\tfrac{1}{2}[\ln |K_\theta| + (\mathbf{Z} - F\hat{\beta})^T K_\theta^{-1} (\mathbf{Z} - F\hat{\beta})] \tag{4}$$

In most examples θ will contain a scale parameter for the covariance, so $c = \kappa c_0$, and $K = \kappa\Sigma$ say. Then

$$L_p(\theta; \mathbf{Z}) = -\tfrac{1}{2}[n \ln \kappa + \ln |\Sigma_\theta| + \frac{1}{\kappa}(\mathbf{Z} - F\hat{\beta})^T \Sigma_\theta^{-1} (\mathbf{Z} - F\hat{\beta})]$$

and hence the MLE of κ is

$$\hat{\kappa} = \frac{1}{n}(\mathbf{Z} - F\hat{\beta})^T \Sigma_\theta^{-1} (\mathbf{Z} - F\hat{\beta}) \tag{5}$$

If $\theta = (\kappa, \phi)$, we find

$$L_p(\phi; \mathbf{Z}) = -\tfrac{1}{2}[n \ln \kappa(\phi) + \ln |\Sigma_\phi| + n] \tag{6}$$

Many authors have advocated maximizing (4) or (6) numerically to find $\hat{\theta}$. The results below show that in general this is nowhere near as easy nor as sensible as has been thought.

2.1 SPATIAL AUTOREGRESSIONS

A conditional autoregression (CAR) can be defined by (Besag, 1975; Ripley, 1981, p. 88)

$$\left.\begin{aligned} &E[Z_i | Z_j, j \neq i] = \mu_i + \sum_{j \neq i} C_{ij}(Z_j - \mu_j) \\ &\text{var}[Z_i | Z_j, j \neq i] = \sigma^2 \end{aligned}\right\} \tag{7}$$

for a symmetric matrix C. Without loss of generality we take $C_{ii} = 0$ for all i. Then this specification defines a multivariate normal distribution with mean $\boldsymbol{\mu}$ and dispersion matrix $K = \sigma^2(I - C)^{-1}$ if and only if $(I - C)$ is strictly positive definite. If we parameterize $\boldsymbol{\mu} = F\beta$ we arrive at (2).

These CAR processes have been quite popular as means of parameterizing the expected correlation between regression errors. In a regular layout such as a lattice, an especially simple form is $C = \phi N$ where $N_{ij} = 1$ if i and j are neighbours, 0 otherwise. On a regular lattice we can show that $(I - \phi N)$ is strictly positive definite if and only if $|\phi|$ is less than the reciprocal of the number of neighbours. (Curiously, no general proof seems available, but the eigenvalues of N are known in the common cases.) In irregular layouts such as economic measurements on local authority areas, a common proposal is $C = \phi W$ where W is a weight matrix and $\phi \geqslant 0$. For example, W_{ij} might be 0 if the areas are not contiguous, and some monotone function of the length of the common boundary if they are contiguous. In such cases we will have a valid model provided ϕ is less than the reciprocal of the largest eigenvalue of W. We have $\Sigma=(I-C)^{-1}$, so

$$\ln|\Sigma|=-\ln|I-\phi W|=-\sum\ln(1-\phi\lambda_i)$$

where (λ_i) are the eigenvalues of W. Since $\operatorname{tr} W=0$, we have $\sum\lambda_i=0$. Suppose the eigenvalues are ordered, with $\lambda_1<0<\lambda_n$. Let $\Lambda=\operatorname{diag}(\lambda_i)$, so $W=C^{\mathrm{T}}\Lambda C$ for an orthogonal matrix C. Then

$$\hat{\sigma}^2=\frac{1}{n}(\mathbf{Z}-F\hat{\beta})^{\mathrm{T}}C^{\mathrm{T}}(I-\phi\Lambda)C(\mathbf{Z}-F\hat{\beta})$$

so

$$\hat{\sigma}^2=\frac{1}{n}\sum_1^n(1-\phi\lambda_i)w_i^2$$

where

$$\mathbf{w}=(w_i)=C(\mathbf{Z}-F\hat{\beta})$$

Thus

$$L_p(\phi)=\text{const}+\tfrac{1}{2}\sum\ln(1-\phi\lambda_i)-\frac{n}{2}\ln\left[\sum(1-\phi\lambda_i)w_i^2\right] \tag{8}$$

This is maximized numerically. Note that $\hat{\beta}$ and hence (w_i^2) depend on ϕ in general.

A similar reduction occurs for the earlier idea of a simultaneous autoregression (SAR) process, defined by

$$\left.\begin{aligned}\mathbf{Z}&=\boldsymbol{\mu}+S(\mathbf{Z}-\boldsymbol{\mu})+\eta\\ \eta&\sim\text{MVN}(\mathbf{0},\sigma^2 I)\end{aligned}\right\} \tag{9}$$

Provided $I-S$ is non-singular this defines a Gaussian process with covariance matrix $\sigma^2(I-S)^{-1}(I-S^{\mathrm{T}})^{-1}$. In practice $S=\phi W$ is chosen as before. It is no longer essential for W to be symmetric, but for simplicity we will assume so. Then

$$\begin{aligned}\ln|\Sigma|&=-2\sum\ln(1-\phi\lambda_i)\\ \hat{\sigma}^2&=\frac{1}{n}\|(I-\phi W)(Z-F\hat{\beta})\|^2\\ &=\frac{1}{n}\sum_1^n(1-\phi\lambda_i)^2w_i^2\end{aligned}$$

so

$$L_p(\phi)=\text{const}+\sum\ln(1-\phi\lambda_i)-\tfrac{1}{2}n\ln\left[\sum(1-\phi\lambda_i)^2w_i^2\right] \tag{10}$$

which is also maximized numerically.

Although both (8) and (10) are functions of a single variable, to maximize them efficiently we do need to know that they are smooth unimodal functions of ϕ. It is clear that they are C^∞ functions. Are they unimodal? The usual way to show this is to prove concavity of L_p at least on $[0, 1/\max \lambda_i)$.

To avoid any complications with $\hat{\beta}(\phi)$, assume that the mean is exactly zero and consider (8). The first term is clearly the sum of concave terms and so is concave, but the second term is convex in ϕ. Their sum can still be concave, as figure 2.1 shows. For small ϕ we have

$$L_p(\phi) \approx \text{const} - \tfrac{1}{2}\phi \sum \lambda_i - \tfrac{1}{4}\phi^2 \sum \lambda_i^2 + \tfrac{1}{2}n[\phi \sum \lambda_i \alpha_i + \tfrac{1}{2}\phi^2 (\sum \lambda_i \alpha_i)^2]$$

where $\alpha_i = w_i^2 / \sum w_j^2$, so

$$L_p''(0) = \tfrac{1}{2}[n(\sum \lambda_i \alpha_i)^2 - \sum \lambda_i^2]$$

If all the α_i are equal, this is negative. However, if only $w_n^2 > 0$, and λ_n is the largest eigenvalue,

$$L_p''(0) = \tfrac{1}{2}(n\lambda_n^2 - \sum \lambda_i^2) > 0$$

Thus $L_p(\;)$ is not always concave, since this situation corresponds to $\mathbf{Z}$ being an eigenvector for λ_n, which can happen. In this extreme case L_p increases with ϕ up to the limit. However, for $\boldsymbol{\alpha}$ near $(0, \ldots, 0, 1)$ $L_p(\;)$ is not concave but has a maximum on $(0, 1/\lambda_n)$, and this event has positive probability. An example is shown in figure 2.2. This means that methods such as Newton's should not be used to maximize L_p.

We can attempt to show unimodality more directly. Consider the first

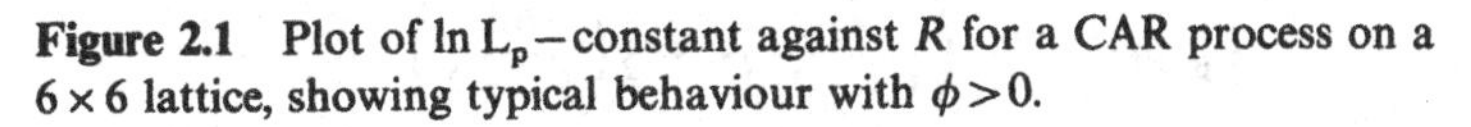

Figure 2.1 Plot of $\ln L_p -$ constant against R for a CAR process on a 6×6 lattice, showing typical behaviour with $\phi > 0$.

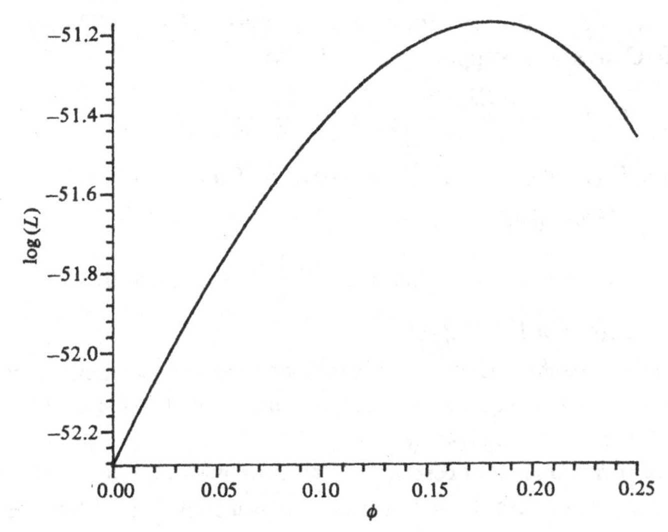

Figure 2.2 Another plot of $\ln L_p$ − constant for a CAR process, this time with $L_p''(0)>0$.

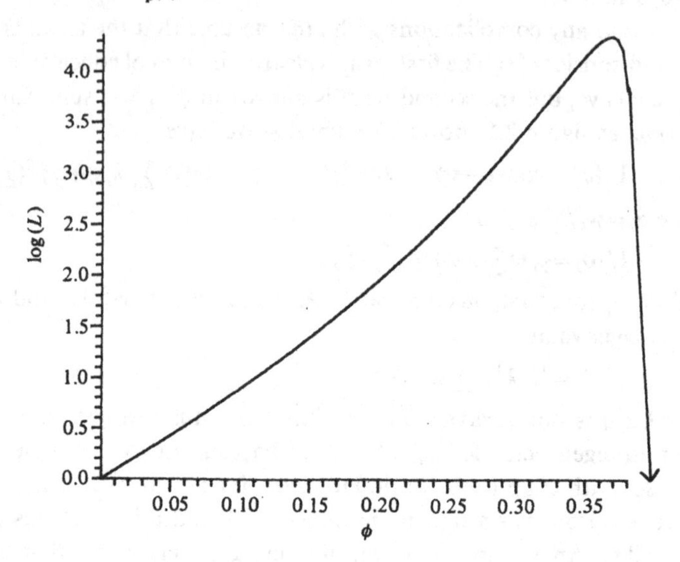

derivative

$$L_p'(\phi)=\tfrac{1}{2}\left[\frac{n\Sigma\lambda_i\alpha_i}{\Sigma(1-\phi\lambda_i)\alpha_i}-\Sigma\frac{\lambda_i}{(1-\phi\lambda_i)}\right]$$

and suppose $n\Sigma\lambda_i\alpha_i \leqslant \Sigma\lambda_i=0$, so $L_p'(0)\leqslant 0$. Then for $\phi>0$

$$L_p'(\phi)\leqslant -\tfrac{1}{2}\Sigma\left(\frac{\lambda_i}{1-\phi\lambda_i}\right)<-\tfrac{1}{2}\left(\frac{n\bar{\lambda}}{1-\phi\bar{\lambda}}\right)=0$$

by Jensen's inequality and the convexity of $x\to x/(1-\phi x)$. Thus in this case $\hat{\phi}=0$. Conversely, suppose $L_p'(0)>0$. Then

$$L_p''(\phi)=\frac{n(\Sigma\lambda_i\alpha_i)^2}{[\Sigma(1-\phi\lambda_i)\alpha_i]^2}-\Sigma\frac{\lambda_i^2}{(1-\phi\lambda_i)^2}$$

Let $a_i=\lambda_i/(1-\phi\lambda_i)$, $A=\Sigma\lambda_i\alpha_i/\Sigma(1-\phi\lambda_i)\alpha_i$. Then

$$L_p'(\phi)=\tfrac{1}{2}[nA-\Sigma a_i]$$

If $L_p'(\phi)<0$, then $\Sigma a_i>nA$, hence $\Sigma a_i^2 \geqslant \frac{1}{n}(\Sigma a_i)^2>nA$, and

$$L_p''(\phi)=nA^2-\Sigma a_i^2<0$$

This suffices to show that $L_p(\ )$ is unimodal unless $\alpha_n=1$, since in that case $L_p(\phi)\to-\infty$ as $\phi\to 1/\lambda_n$ and hence at some point $L_p'(\phi)<0$. Beyond that point $L_p(\ )$ decreases steadily.

The conclusion is that certain optimization methods will work correctly (for example, success–failure searches with interpolation) but some may fail. When a parameterized mean function is present we do not know even if $L_p(\phi)$ is unimodal.

An SAR process for small ϕ is essentially the same as a CAR process with half the value of ϕ, so concavity also fails with positive probability in this case. We have

$$L_p'(\phi) = \frac{n\Sigma(1-\phi\lambda_i)\lambda_i\alpha_i}{(1-\phi\lambda_i)^2\alpha_i} - \Sigma\left(\frac{\lambda_i}{1-\phi\lambda_i}\right)$$

$$L_p''(\phi) = \frac{2n[\Sigma(1-\phi\lambda_i)\lambda_i\alpha_i]^2}{[\Sigma(1-\phi\lambda_i)^2\alpha_i]^2} - \frac{n\Sigma\lambda_i^2\alpha_i}{\Sigma(1-\phi\lambda_i)^2\alpha_i} - \Sigma\left(\frac{\lambda_i}{1-\phi\lambda_i}\right)^2$$

As above, when $L_p'(\phi) < 0$,

$$\Sigma\left(\frac{\lambda_i}{1-\phi\lambda_i}\right)^2 > \frac{n[\Sigma(1-\phi\lambda_i)\lambda_i\alpha_i]^2}{[\Sigma(1-\phi\lambda_i)^2\alpha_i]^2}$$

so

$$L_p''(\phi) < \frac{2n[(\Sigma\lambda_i\alpha_i)^2 - \Sigma\lambda_i^2\alpha_i]}{[\Sigma(1-\phi\lambda_i)^2\alpha_i]^2} \leqslant 0$$

and we conclude that L_p is unimodal. The case of a parameterized mean function is again unknown.

2.2 RANGE PARAMETERS

An alternative parameterization for the covariance function $c(\mathbf{x}, \mathbf{y})$ is for it to decay smoothly with the distance between the two points,

$$c(\mathbf{x}, \mathbf{y}) = c(\|\mathbf{x} - \mathbf{y}\|) \qquad \textbf{(11)}$$

Some suggestions are

$$c(r) = \sigma^2 \exp\{-(r/R)\} \qquad \textbf{(12a)}$$

$$c(r) = \sigma^2 \exp\{-(r/R)^2\} \qquad \textbf{(12b)}$$

$$c(r) = \sigma^2 \gamma_B(r)/\gamma_B(0) \qquad \textbf{(12c)}$$

where $\gamma_B(r)$ is the area of overlap of discs B of diameter R with centres r apart. In each case R is a range parameter. These choices have been used in the prediction problem (Ripley, 1981, chapter 4) and in regression with spatially dependent errors (Cook and Pocock, 1983).

Our first example is the 52-point dataset on topographic heights used by Ripley (1981, pp. 58–72) to illustrate universal kriging. There (11) and (12a) were suggested with $R \approx 2$ and $\sigma \approx 65$ by comparing theoretical curves with an empirical correlogram. Figures 2.3 and 2.4 show some likelihood profiles when $f(\mathbf{x}) \equiv 1$. These are far from unimodal!

(It might be thought that the multiple maxima could be due to rounding errors. We checked this by recomputing the figures on a CRAY-XMP in

Figure 2.3 Profile likelihood for R for the topographic data.

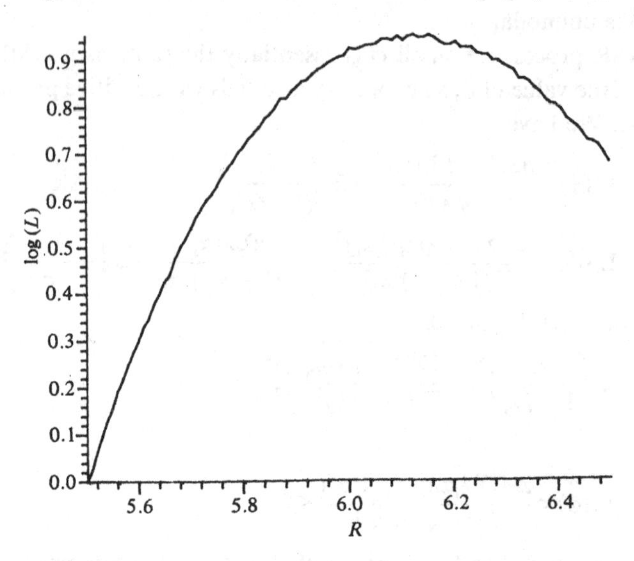

Figure 2.4 Likelihood profile for (σ, R) for the topographic data with an unknown overall mean. (Adapted with permission from Warnes and Ripley, 1987.)

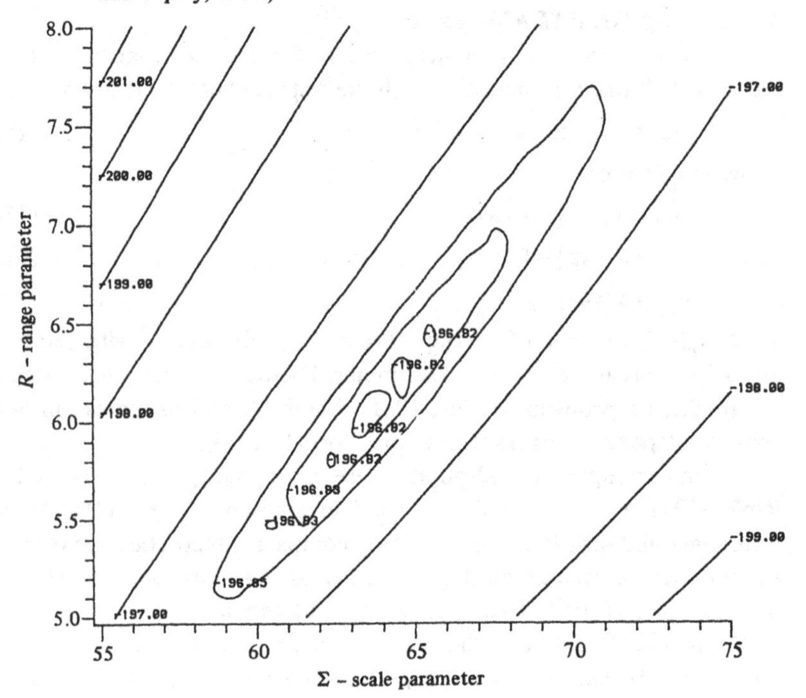

double the precision of a VAX 8600, with negligible change in the results. The maxima are also *not* due to the contouring algorithm, which uses only linear interpolation and so cannot introduce false maxima. It remains possible that the multiple maxima are numerical.)

A further worrying point is that none of the local maxima appear at all consistent with the empirical correlogram (figure 2.5) for these data.

An obvious suggestion is that these problems arise because the model is inappropriate. To check this we simulated data. Following Mardia and Marshall (1984) we chose model (12c). They used the function appropriate to $\mathbb{R}^3$

$$c(r)=\begin{cases}\sigma^2\left[1-\dfrac{3}{2}\left(\dfrac{r}{R}\right)+\dfrac{1}{2}\left(\dfrac{r}{R}\right)^3\right] & r\leqslant R\\ 0 & r\geqslant R\end{cases}$$

which is also a valid covariance function in $\mathbb{R}^2$. Mardia and Marshall reported good agreement between their simulation results for this model and some asymptotic theory. Figures 2.7 and 2.8 show profile likelihoods for simulations of this process on a 6×6 and 8×8 grid respectively. These are typical of the behaviour of a number of simulations. Multimodality

Figure 2.5 Empirical correlogram for the topographic data with covariance functions fitted by eye. (This and figure 2.6 are reprinted by permission of John Wiley & Sons, Inc. from B. D. Ripley, *Spatial Statistics*, copyright © John Wiley & Sons, Inc. 1981.)

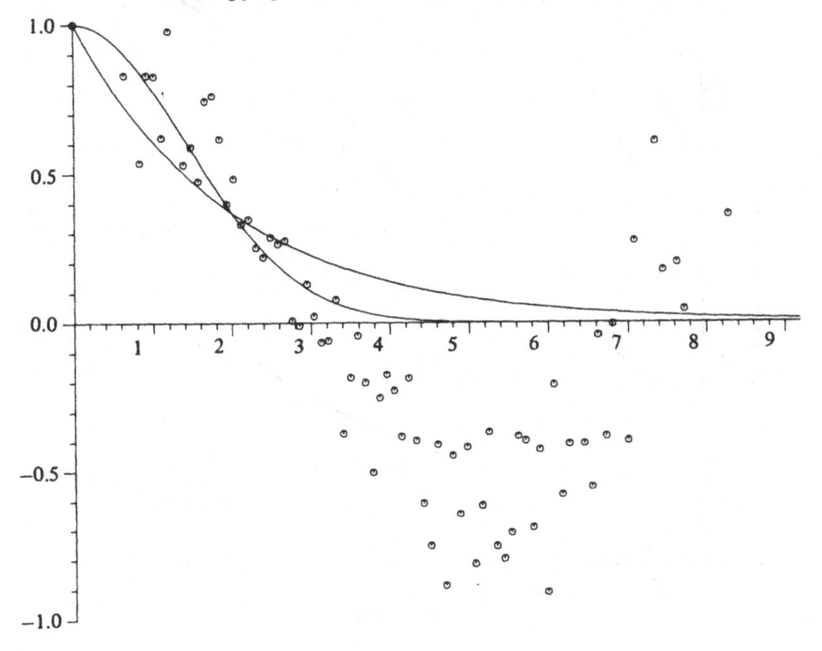

Figure 2.6 As figure 2.5 for residuals from a quadratic trend surface.

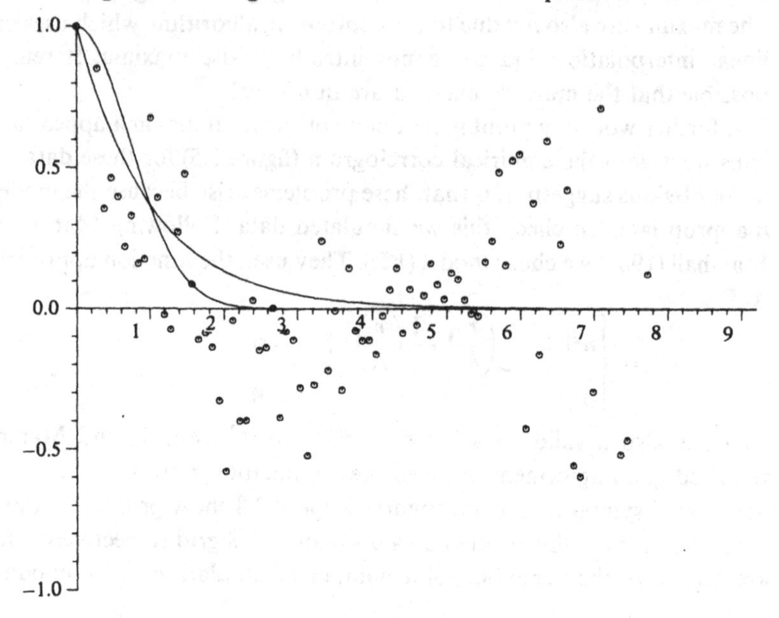

Figure 2.7 Profile log likelihood for (σ, R) from a simulation of (12c) on a 6×6 grid. The search path shown is that of Fisher scoring.

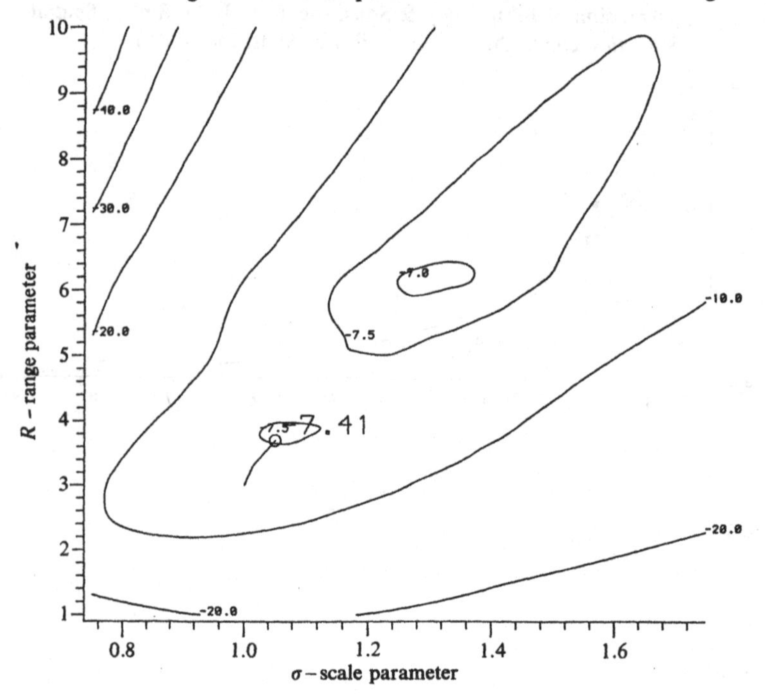

Figure 2.8 As for figure 2.7 but for an 8 × 8 grid. Note that Fisher scoring oscillates.

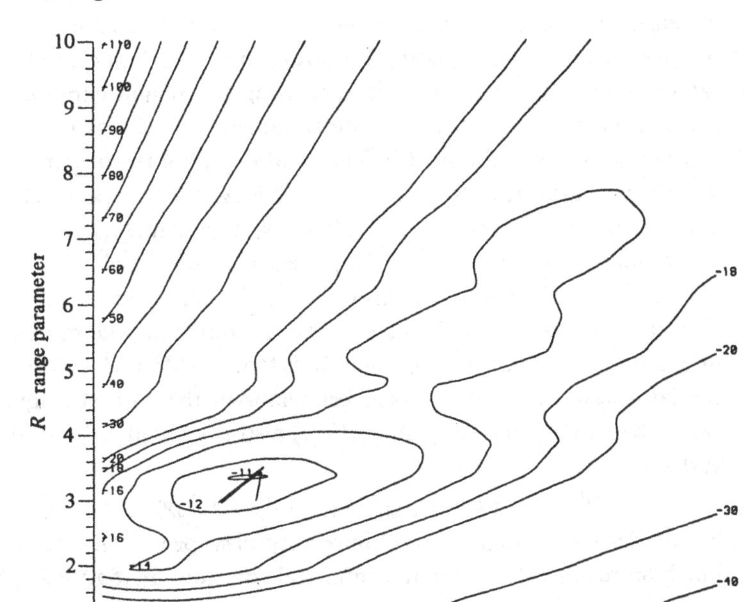

was quite common, and the global maximum was often well away from the true value of (σ^2, R) of (1, 3).

These surfaces shown in figures 2.4–2.8 will be rather difficult to maximize numerically other than by a grid search. Most optimization algorithms expect at least unimodality. Mardia and Marshall used Fisher scoring, which is especially dangerous on complex surfaces. Figure 2.8 shows how it can oscillate. Since Mardia and Marshall's simulation results were for local maximizations starting *from the true value*, they are not representative of the actual maximum likelihood estimators.

2.3 DISCUSSION

These examples show that, in some cases at least, the profile likelihood can be a very inappropriate summary of the information about covariance parameters in a dataset. It is not clear why this is so. Mardia and Marshall (1984) gave some asymptotic theory that gave the expected results of strong consistency, best asymptotic normality and so on. However, we need to consider what is meant by asymptotic here. We could either add further sampling points $\mathbf{x}_i$ near the existing points (corresponding to more intense exploration of a bounded area) *or* we could explore an increasing region at about the same intensity (and there are compromises between these two alternatives). Mardia and Marshall's conditions are rather technical, but amount to increasing the region explored. As the correlation given by (11) is of short range, we will essentially get an increasing number of almost independent copies of the original problem. Thus the classical results are not surprising. As a minor point, (12c) is not covered by their theory (it is not C^4) and so their simulation results could not validate a theory of which they are not an example!

Checking the individual terms in the profile likelihood $L_p(\sigma^2, R)$ shows that each is well behaved but that the multimodality comes from their interaction. One plausible explanation is that the MLE $\hat{R}$ is responding to the large-scale rather than local behaviour of the dataset. Figure 2.3 is rather basin-shaped, and Ripley (1981) also suggested a quadratic trend surface

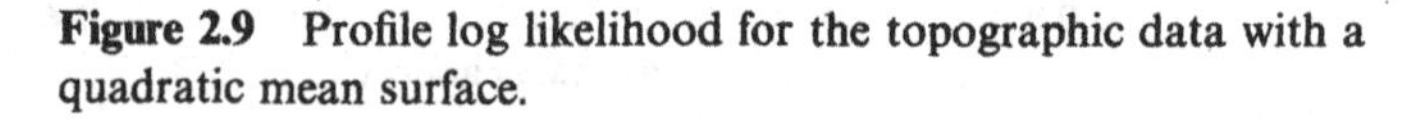

$$f(\mathbf{x})^{\mathrm{T}}\beta = \beta_1 + \beta_2 x + \beta_3 y + \beta_4 x^2 + \beta_5 xy + \beta_6 y^2$$

Figure 2.9 shows that the multimodality disappears. The value of $\hat{R}$ is much more consistent with the empirical correlogram (figure 2.6) but still seems to be an overestimate.

Figure 2.9 Profile log likelihood for the topographic data with a quadratic mean surface.

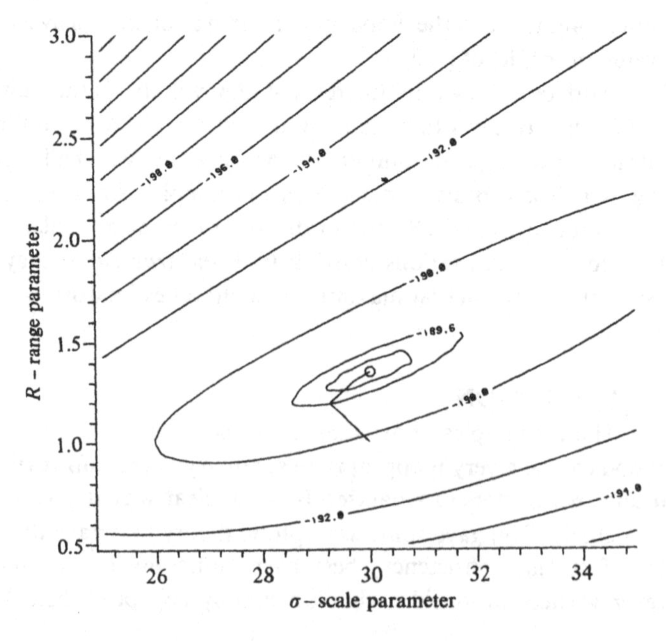

These examples also cause difficulties for Bayesian estimation. It is plausible from the figures (and could be confirmed by careful numerical integration) that for any reasonably diffuse prior on (σ, R, β) the posterior density will have its mass concentrated on unrealistic values in the topographic data example, and that this is true for the posterior distribution for R in all cases. Suspicion concentrates on the information in the likelihood surface about σ, for with σ known the support for R concentrates on the 'correct' values.

The position at present is still unsatisfactory. The results of this chapter present a new and completely unexpected failure of likelihood estimation. They have considerable ramifications for theories of statistical inference. Further, we still need efficient parameter estimators for spatially parameterized covariance functions, and work is in progress on alternative approaches.

3

Edge corrections for spatial point processes

The set-up for this chapter is a spatial point process defined throughout $\mathbb{R}^d$ but observed only within a bounded window E. The point process will be stationary under translation (at least). Formally, a realization of a point process consists of an infinite number of points without accumulation points, so there is a finite number of points in E. Multiple points are not allowed, so that each point of $\mathbb{R}^d$ occurs at most once. Various measure-theoretic descriptions are possible (Ripley, 1976a) but all are isomorphic to that describing the point process by $\{N(A) \mid A \text{ bounded Borel}\}$ where $N(A)$ denotes the number of points within A. We will use this form.

The first moment measure $\Lambda(A) = \mathrm{E}N(A)$ gives the expected number of points in A. By stationarity it must be proportional to Lebesgue measure ν, so

$$\mathrm{E}N(A) = \lambda\nu(A)$$

for a constant λ known as the *intensity*. The second moment measure is defined by

$$\mu_2(A \times B) = \mathrm{E}[N(A)N(B)]$$

and this can be reduced to a simpler form under stationarity (Ripley, 1976b). Under the additional assumption of *isotropy* we find

$$\mathrm{E}N(A)N(B) = \lambda\nu(A \cap B) + \lambda^2 \int_0^\infty \nu_t(A \times B)\,\mathrm{d}K(t) \tag{1}$$

and

$$\nu_t(A \times B) = \int \sigma_t(\{y - x \mid x \in A,\ y \in B,\ d(x, y) = t\})\,\mathrm{d}\nu(x)$$

where σ_t is the uniform probability on the surface of the sphere in $\mathbb{R}^d$ of radius t. The crucial part of (1) is that the second-moment description is reduced to an increasing function $K(\ )$ on $(0, \infty)$. Taking $K(0)=0$ we find

$$\lambda K(t) = \mathrm{E}[\text{number of points with distance} \leqslant t \,|\, \text{point at } \mathbf{x}] \tag{2}$$

using Palm probabilities to formalize the conditioning on an event of probability zero (Ripley, 1977, p. 190). Thus K is a sort of interpoint distance distribution function, although $K(t) \to \infty$ as $t \to \infty$.

It must be stressed that since point processes are not Gaussian processes, they are not determined by their first two moment measures. Although this was pointed out in the author's early work, it has been overlooked in many of the applications, and Baddeley and Silverman provided a warning example (1984). It is a measure of the success of second-order methods that they have been mistakenly assumed to be all-powerful!

Another family of descriptions is

$$\mathrm{P}[N(A_1)=0, \ldots, N(A_m)=0]$$

for (disjoint) sets A_i. The simplest member of this family is

$$\mathrm{P}[N(b(\mathbf{0}, t))=0]$$

and estimates of this have been termed 'empty-space' statistics by Lotwick and Silverman (1982). The original application in Ripley (1977) used

$$p(t) = \mathrm{P}[N(b(\mathbf{0}, t))>0] \tag{3}$$

essentially the same function, but measuring occupied rather than empty space.

Finally, a rather incomplete family of descriptions is based on nearest neighbour distances d_i, this being the distance from the ith observed point in E to its nearest neighbour. We can consider the c.d.f. of d_i, or the distribution of $T=d_1+\cdots+d_n$. Such ideas have been proposed by Diggle (1979, 1983) but are less widely used, for reasons which will emerge below.

The edge effect problem will be severe when E has a complex shape. An extreme case is the nesting pattern of birds within woodland considered by Ripley (1985). In the past users (including the author) have deliberately chosen a simple window E (a circle, square or rectangle) either to reduce edge effects or to ease the computation of edge corrections. With modern computing equipment this is no longer necessary, and it is feasible to consider essentially arbitrary windows E, certainly as complicated as a map of the islands of Great Britain and Ireland.

There are two quite distinct strategies evident in correcting for edge effects. The first is to compute their magnitude and to rescale the statistics accordingly. For example, Brown and Rothery (1978) compute statistics based on (d_i^2). For points near the edge the corresponding d_i^2 will be inflated (the nearest point possibly lies outside E), and the effect of this on the mean and variance of the final statistic can be computed (perhaps by simulation). The problem is that the computation has to be done for every E and every point process model. The alternative strategy is to attempt to compensate directly for the edge effects. Examples will be seen in §3.2. If such a scheme is possible it leads to much neater and more universal solutions, and hence is to be preferred.

A final solution to the problem of edge effects is to abolish edges! Spaces such as the sphere and the torus have no edges and hence no edge effects. For data within a rectangular window a common practice is to wrap the window onto a torus. Of course this does not abolish edge *effects*, since one has still to consider exact results on a torus as approximate results on the plane. This is conveniently ignored.

3.1 EDGE CORRECTIONS FOR NEAREST NEIGHBOUR METHODS

We consider both methods based on (d_i) and on $p(\ )$ as nearest neighbour methods, for $p(\ )$ is the c.d.f. of the distance from the origin to the nearest point (directly from the definition (3)). Similar edge corrections can be applied to each.

One way of considering the nearest neighbour distance d is to view (in $\mathbb{R}^2$) πd^2 as the 'area swept out' in a spiral search for the nearest point. For a Poisson process it is πd^2, an unbiased estimator of $1/\lambda$. If the edge intervenes in the search, the actual area searched (now less than πd^2) is still unbiased for $1/\lambda$. Thus one could compute an 'equivalent nearest neighbour distance' as $\sqrt{(\text{area swept out}/\pi)}$. This method of edge correction has several disadvantages. It is not particularly easy to compute for complex windows E. It gives statistics whose distribution depends quite strongly on the shape of E even for a Poisson process and its variability is almost impossible to calculate. Indeed, edge correction merely exacerbates a major problem with nearest neighbour methods, that not even their null-hypothesis distributions can be found analytically, and one is reduced to simulation results.

About the only method based on (d_i) whose properties are well understood is the Clark–Evans (1954) statistic $T=\Sigma d_i$. Its originators claimed that it was normally distributed with a mean and variance which ignored edge effects *and* dependence between the distances d_i. The problem was investigated by several authors in the 1970s (Matérn, 1972; Persson, 1972; De Vos, 1973; Hsu and Mason, 1974; Vincent *et al.*, 1976; Donnelly, 1978). Some of these papers, especially Donnelly's, managed to calculate $\mathrm{E}\,d_i$ and hence $\mathrm{E}\,T$ under assumptions of straight boundaries, and, eventually, sharp corners. This provides the most comprehensive example of our first type of correction, for the statistic used is still

$$\mathrm{CE}=\frac{T-\mathrm{E}\,T}{\sqrt{\operatorname{var} T}}$$

but $\mathrm{E}\,T$ and var T are (approximately) corrected for edge effects and dependence. Donnelly demonstrated approximate normality of CE from simulations, but although central limit theorems have been sketched, to my knowledge no complete proof of asymptotic normality has ever been published. This is a good example of how edge effects complicate the calculations.

Another method is available when estimating a c.d.f. such as $p(\ )$. The *border* method has quite wide applicability, and we will meet it again in §3.2. Select a set $E_t \subset E$ such that $E_t \oplus b(\mathbf{0}, t) \subset E$. Then provided we consider only distances from points in E_t which are shorter than t, we can ensure that all possible points are in E and hence observed. In the case of the c.d.f. of nearest neighbour distances we will know whether or not $d_i \leqslant t$ for all points $\mathbf{x}_i \in E_t$. Thus an edge-corrected estimate of the c.d.f. G of (d_i) might be

$$\frac{\#(\text{points with } d_i \leqslant t \text{ in } E_t)}{\#(\text{points in } E_t)}$$

One can choose E_t in a slightly less restrictive way. All we actually need is that $\mathrm{b}(\mathbf{x}_i, t) \subset E$ for all $\mathbf{x}_i \in E_t$. Define $r_i = \min\{d(\mathbf{x}_i, \mathbf{x}) \mid \mathbf{x} \notin E\}$. An improved estimator is

$$\hat{G}(t)=\frac{\#(\text{points with } d_i \leqslant t \leqslant r_i)}{\#(\text{points with } t \leqslant r_i)} \tag{4}$$

This can also be applied to estimate $p(\ )$. By stationarity,

$$p(t)=\mathrm{P}(\exists \text{ a point within distance } t \text{ of } \mathbf{x})$$

for any point $\mathbf{x}$. Sample m points $\mathbf{y}_1, \ldots, \mathbf{y}_m \in E$ and calculate

$$d_i=\min\{d(\mathbf{x}_j, \mathbf{y}_i) \mid \mathbf{x}_j \text{ point of pattern}\}$$

Let

$$\hat{p}(t)=\frac{\#\{i \mid d_i \leqslant t \leqslant r_i\}}{\#\{i \mid t \leqslant r_i\}} \tag{5}$$

where r_i is the distance from $\mathbf{y}_i$ to E^{c}, as before. Now $\hat{p}(t)$ is an unbiased estimator of $p(\ )$ whatever the underlying (stationary) point process, for

$$\begin{aligned}\#\{i \mid r_i \geqslant t\}\mathrm{E}\hat{p}(t) &= \sum_{i: r_i \geqslant t} \mathrm{P}(d_i \leqslant t)\\ &= \sum_{i: r_i \geqslant t} \mathrm{P}(N[b(\mathbf{y}_i, t)] > 0)\\ &= \sum_{i: r_i \geqslant t} p(t)\end{aligned}$$

Although unbiased, $\hat{p}(t)$ does have some disadvantages. Note that $p(\ )$ is monotone but $\hat{p}(\ )$ need not be (figure 3.1). Further,

$$\{\mathbf{y}_i \mid r_i \geqslant t\} = \{\mathbf{y}_i\} \cap [E \ominus b(\mathbf{0}, t)]$$

and so for large t we will find few (or no) points meeting the condition. Define

$$\begin{aligned} t_{\mathrm{m}} &= \min\{t \mid E \ominus b(\mathbf{0}, t) = \varnothing\} \\ &= \min\{t \mid \nexists \mathbf{x} \in E \text{ with } b(\mathbf{x}, t) \subset E\} \end{aligned}$$

Clearly for $t \geqslant t_{\mathrm{m}}$ we are unable to estimate $p(t)$; for points $\mathbf{y}$ we may know that $d_i \leqslant t$ but can never know that $d_i > t$. Thus without further knowledge of the process we cannot estimate $\hat{p}(t)$ for $t \geqslant t_{\mathrm{m}}$ in an unbiased way. Fortunately, for regularly shaped domains we will usually find $p(t_{\mathrm{m}}) \approx 1$. Suppose we observe n points in E, and E has area a. Then for a Poisson process

$$p(t) = 1 - \exp\{-\lambda \pi t^2\}$$

which is essentially one for $\lambda \pi t^2 > 4.6$ or $t > 1.21/\sqrt{\lambda}$. As the usual estimator of λ is n/a, we find

$$\frac{t}{\sqrt{a}} > \frac{1.2}{\sqrt{n}}$$

In the most extreme possible case of regular packing on a triangular lattice, $p(t) = 1$ for $t > \sqrt{[(s\sqrt{2}a)/48n]}$ or

$$\frac{t}{\sqrt{a}} > \frac{0.38}{\sqrt{n}}$$

At the other extreme we would have all points in a single cluster and so can give no sensible bound. What we can conclude from these calculations is that unless n is very small *or* the pattern is very clustered then $\hat{p}(\ )$ will approach one well before t_m.

Thus far we have not considered the variability of $\hat{p}(\ )$. This is discussed in Ripley (1981, p. 151) where a systematic layout of $\{\mathbf{y}_i\}$ is shown to be optimal. It is clear that we would expect $\hat{p}(t)$ to be more variable for large t, when the implied sum in (5) is over a smaller number of terms. In practice it is for large t that monotonicity breaks down, as shown in Figure 3.1.

The unbiasedness of $\hat{p}(t)$ for $t \in (0, t_{\mathrm{m}})$ does not apply to $\hat{G}(t)$, because the denominator of (4) is random whereas that of (5) is fixed. We might expect $\hat{G}(\ )$ to be approximately unbiased. (Note that Diggle, 1983, p. 73, incorrectly assumes $\hat{G}(\ )$ to be exactly unbiased.) However, lack of unbiasedness here is not too serious a disadvantage, because $G(\ )$ is unknown for almost any model except a Poisson process. As it has to be

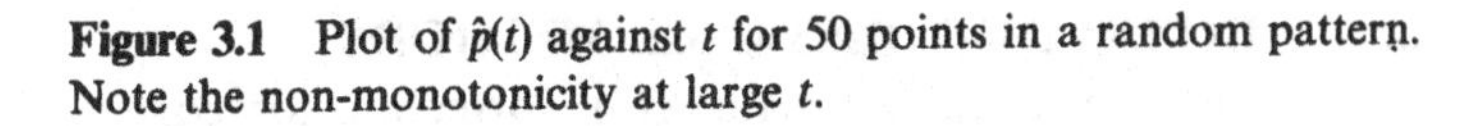

Figure 3.1 Plot of $\hat{p}(t)$ against t for 50 points in a random pattern. Note the non-monotonicity at large t.

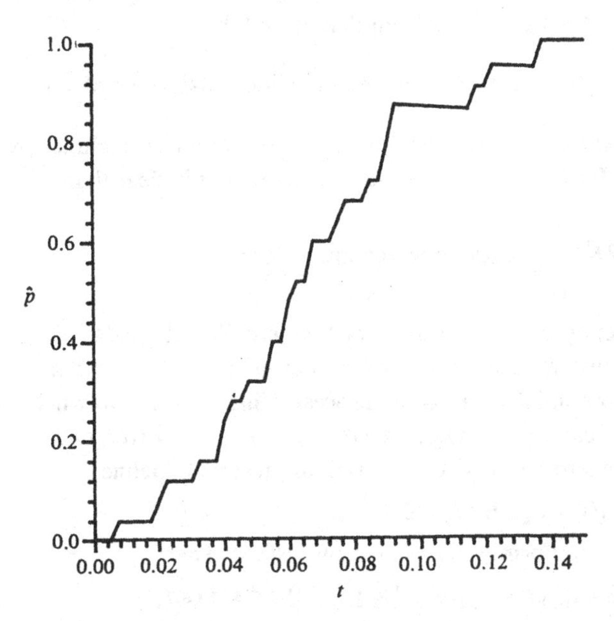

estimated by simulation, we may as well redefine $G(t) = \mathrm{E}\hat{G}(t)$!

Hanisch (1984) proposed a modification of $\hat{G}(\ )$ which is slightly more efficient. Instead of considering points $\mathbf{x}_i$ with $r_i \geqslant t$, consider points with $d_i \leqslant r_i$ so their nearest neighbours in $\mathbb{R}^d$ are known to be within E. Then

$$\hat{G}_1(t) = \frac{\#(\text{points with } d_i \leqslant r_i,\ t)}{\#(\text{points with } d_i \leqslant r_i)}$$

is approximately unbiased, and has the additional advantage of being monotone, unlike $\hat{G}(t)$. The same idea can be used for $p(t)$, but exact unbiasedness is lost there.

A serious problem with both $\hat{p}(\ )$ and $\hat{G}(\ )$ is to approximate their sampling distribution. Baddeley (1980) gives some asymptotic theory for $\hat{p}(\ )$ for a Poisson point process. Results for nearest neighbour distances intrinsically rescale so that the mean distance is constant, hence there are no problems here with the asymptotic sense. This rescaling ensures that edge effects are asymptotically negligible. As Baddeley points out, his results are less reliable for values of t with $p(t)$ near one, exactly where edge effects are most important.

No distribution theory is known for $\hat{G}(\ )$. Diggle, its main exponent, used Monte-Carlo methods (see chapter 4).

3.2 INTERPOINT DISTANCE METHODS

These methods estimate $K(\ )$, correcting for edge effects. The simplest estimator comes from definition (2):

$$\hat{K}_0(t) = \frac{1}{\lambda n} \sum_1^n \#(\text{other points within distance } t \text{ of } \mathbf{x}_i)$$

This is clearly unbiased provided all other points are included, whether in E or not. If only observed points are recorded it is clear that $\mathrm{E}\,\hat{K}_0(t) < K(t)$. In fact

$$\frac{\lambda \hat{K}_0(t)}{(n-1)} = \frac{\#(\text{pairs which are } t\text{-close})}{\#(\text{pairs})}$$

and the expectation of this is $F(t)$, where F is the c.d.f. of the distance between two randomly located points in E, provided we are observing either a binomial or Poisson process. This c.d.f. is known exactly in a number of cases. The magnitude of the bias (i.e. the divergence of $F(\ )$ from $K(\ )$) seems to be insufficiently well appreciated. Define

$$\gamma_E(s) = \mathrm{E}[\nu(E \cap E + Z)]$$

where Z is uniformly distributed on $\partial b(0, s)$. Then

$$\begin{aligned} \mathrm{E}\,\lambda n \hat{K}_0(t) &= \mu_2\{(\mathbf{x}, \mathbf{y}) \mid \mathbf{x}, \mathbf{y} \in E, 0 < d(\mathbf{x}, \mathbf{y}) \leqslant t\} \\ &= \lambda^2 \int_0^t \nu_s(E \times E) \mathrm{d}K(s) \\ &= \lambda^2 \int_0^t \gamma_E(s) \mathrm{d}K(s) \end{aligned}$$

so

$$\mathrm{E}\, n\hat{K}_0(t) = \lambda \int_0^t \gamma_E(s) \mathrm{d}K(s) \tag{6}$$

For a binomial process we saw that

$$\mathrm{E}\, n\hat{K}_0(t) = n(n-1)F(t)/\lambda$$

so for a Poisson process

$$\mathrm{E}\, n\hat{K}_0(t) = (\lambda a)^2 F(t)/\lambda = \lambda a^2 F(t)$$

and we can identify

$$F(t) = \int_0^t \gamma_E(s) 2\pi s \, \mathrm{d}s / a^2 \tag{7}$$

For simple shapes $\gamma_E(\ )$ can be computed. For example, for a disc of radius R

$$\begin{aligned} \gamma_E(s) &= a - 2R^2 \sin^{-1} t - 2R^2 t \sqrt{(1+t^2)}, \quad t = s/2R \\ &\approx a - us/\pi - s^3/6R \end{aligned} \tag{8a}$$

where u is the perimeter length of E. For a rectangle

$$\gamma_E(s)=a-us/\pi+s^2/\pi \tag{8b}$$

provided s does not exceed the length of the shortest side. This suggests the general approximation

$$\gamma_E(s)\approx a-us/\pi \text{ for small } s \tag{8}$$

which is confirmed by the general result that $\gamma'_E(0)=-u/\pi$ (Matheron, 1975, §4.3) for convex E. Combining (7) and (8) gives

$$F(t)\approx\left(\frac{\pi t^2}{a}-\frac{2ut^3}{3a^2}\right)$$

Note that the approximation for γ_E and hence for F will be particularly good for a disc. Figure 3.2 illustrates these approximations. (It seems that the relative error is $O(t^3)$ for a convex body E with a smooth boundary, from work of Geciauskas.)

Now return to the general formula (6). Clearly $\gamma_E(s)<a$ for $s>0$, so $\mathrm{E}\,n\hat{K}_0(t)<\lambda aK(t)$. Since $\mathrm{E}\,n=\lambda a$,

$$\mathrm{E}\,\hat{K}_0(t)\approx\frac{\mathrm{E}\,n\hat{K}_0(t)}{\mathrm{E}\,n}=\frac{\lambda\int_0^t\gamma_E(s)\mathrm{d}K(s)}{\lambda a}<K(t)$$

A more careful analysis of the approximation here will show a further bias since $\hat{K}_0(t)$ will be positively correlated with n (if they are many points, more are likely to be close by). We can check this exactly for a Poisson

Figure 3.2 Approximations for $\gamma_E(t)$. (a) Circle radius 2. (b) Unit square. In each case the solid line is exact, the dashed lines are (8). In (a) the dotted line (almost coincident with the solid line) is (8a).

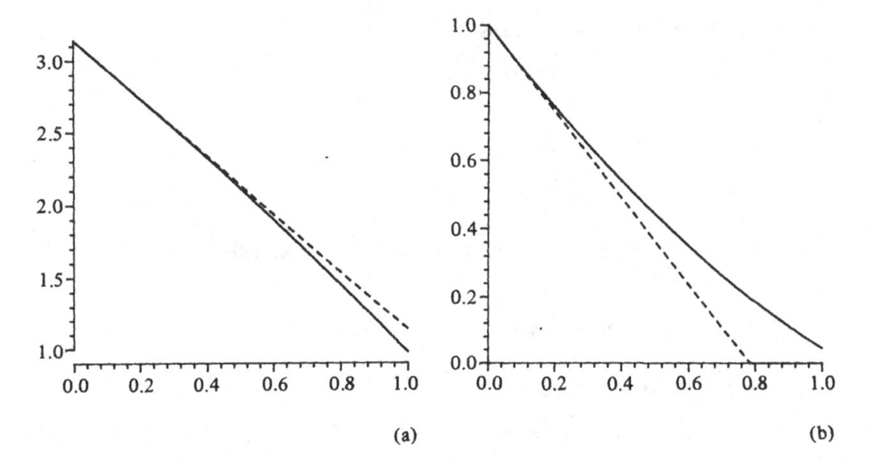

process. There

$$E[\hat{K}_0(t)\,|\,n] = \frac{(n-1)F(t)}{\lambda}$$

so

$$\mathrm{E}\,\hat{K}_0(t) = \frac{(\lambda a - 1)}{\lambda}\,F(t) \approx \pi t^2 - \frac{2ut^3}{a} - \frac{\pi t^2}{\lambda a}$$

and both bias terms are downwards. This is true in general.

Some general results will help us to calculate the variance of $\hat{K}_0(t)$. Note that $n\hat{K}_0(t)$ is of the form

$$T = \sum_{\substack{\text{sample points}\\ \mathbf{x} \neq \mathbf{y}}} \phi(\mathbf{x}, \mathbf{y})$$

for a symmetric function ϕ. Then, using factorial moment measures α_r (Daley and Vere-Jones, 1972),

$$\mathrm{E}\,T = \int_{E^2} \phi(\mathbf{x}, \mathbf{y})\mathrm{d}\alpha_2(\mathbf{x}, \mathbf{y}) \tag{9}$$

$$\begin{aligned}\mathrm{E}\,T^2 = {} & \int_{E^4} \phi(\mathbf{x}, \mathbf{y})\phi(\mathbf{u}, \mathbf{v})\mathrm{d}\alpha_4(\mathbf{x}, \mathbf{y}, \mathbf{u}, \mathbf{v}) \\ & + 4\int_{E^3} \phi(\mathbf{x}, \mathbf{y})\phi(\mathbf{x}, \mathbf{z})\mathrm{d}\alpha_3(\mathbf{x}, \mathbf{y}, \mathbf{z}) \\ & + 2\int_{E^2} \phi(\mathbf{x}, \mathbf{y})^2\mathrm{d}\alpha_2(\mathbf{x}, \mathbf{y})\end{aligned}$$

Let ν_d denote Lebesgue measure on $\mathbb{R}^d$. Then for a Poisson process

$$\alpha_r = \lambda^r \nu_{2r}$$

whereas for a binomial process

$$\alpha_r = n(n-1)\ldots(n-r+1)a^{-r}\nu_{2r}$$

Define

$$S = \int_{E^2} \phi(\mathbf{x}, \mathbf{y})\mathrm{d}\mathbf{x}\mathrm{d}\mathbf{y}$$

$$S_1 = \int_{E^3} \phi(\mathbf{x}, \mathbf{y})\phi(\mathbf{x}, \mathbf{z})\mathrm{d}\mathbf{x}\mathrm{d}\mathbf{y}\mathrm{d}\mathbf{z} = \int_E \left\{\int_E \phi(\mathbf{x}, \mathbf{y})\mathrm{d}\mathbf{y}\right\}^2 \mathrm{d}\mathbf{x}$$

$$S_2 = \int_{E^2} \phi(\mathbf{x}, \mathbf{y})^2\mathrm{d}\mathbf{x}\mathrm{d}\mathbf{y}$$

Then for a Poisson process

$$\mathrm{var}_\lambda T = 4\lambda^3 S_1 + 2\lambda^2 S_2 \tag{10}$$

and for a binomial process, with $\mathbf{X}_i$ uniformly distributed within E,

$$\begin{aligned}\mathrm{var}_n T &= 4n(n-1)(n-2)a^{-3}S_1 - n(n-1)(4n-6)a^{-4}S^2 \\ &\quad + 2n(n-1)a^{-2}S_2 \\ &= 2n(n-1)[\mathrm{var}\,\phi(\mathbf{X}_1, \mathbf{X}_2) + 2(n-2)\mathrm{var}\{\mathrm{E}[\phi(\mathbf{X}_1, \mathbf{X}_2)|\mathbf{X}_2]\}]\end{aligned} \tag{11}$$

Let us apply these results to $\lambda n\hat{K}_0(t)$. Then $\phi(\mathbf{x}, \mathbf{y}) = 1(d(\mathbf{x}, \mathbf{y}) \leqslant t)$, so

$$S_2 = S = \int_0^t \gamma_E(s)\mathrm{d}s \approx \pi a t^2 - 2ut^3/3$$

Further, for small t, $\int \phi(\mathbf{x}, \mathbf{y})\mathrm{d}\mathbf{y} = \pi t^2$ except for $\mathbf{x}$ near the boundary, so $S_1 \approx a\pi^2 t^4$. Thus to first order

$$\mathrm{var}_\lambda[n\hat{K}_0(t)] \approx 4\lambda a\pi^2 t^4 + 2a\pi t^2$$

and

$$\mathrm{var}_n[n\hat{K}_0(t)] \approx 2n(n-1)\pi t^2/\lambda^2 a \approx 2\pi a t^2$$

since $\lambda = n/a$ for a binomial process. The result for $n\hat{K}_0(t)$ is a little misleading, since it includes the variability of $\mathrm{E}[n\hat{K}_0(t)\,|\,n]$ in its first term. For a Poisson process λ is unknown, so we must estimate it by $\hat{\lambda} = n/a$. Then

$$\mathrm{E}[\hat{K}_0(t)|n] = \frac{(n-1)a}{n}\,F(t) \approx \pi t^2 - \frac{2ut^3}{a}$$

$$\mathrm{var}[\hat{K}_0(t)|n] \approx \frac{2\pi a t^2}{n^2}$$

Thus we have (ignoring the fact that $\mathrm{E}[1/n^2]$ is infinite for the present)

$$\mathrm{E}_\lambda \hat{K}_0(t) \approx \pi t^2 - \frac{2ut^3}{a}$$

$$\mathrm{var}_\lambda \hat{K}_0(t) \approx \frac{2\pi t^2}{\lambda^2 a}$$

and we obtain more accurate approximations in §3.3.

Edge corrections

Our analysis thus far has uncovered two problems with the naive estimator $\hat{K}_0(t)$. The bias for small λ is unavoidable; it stems from the need for two points to compute *any* interpoint distance information. The bias from edge correction is both more serious and avoidable. There are several essentially different ideas for doing so, as well as variants on most of them.

(a) Toroidal edge correction for a rectangle. For a torus $u=0$ so to first order $\gamma_E(s) = a$. In fact, of course, $\gamma_E \equiv a$ from the definition. For a process which is defined on a torus there are no edge effects to be corrected. For any process other than a Poisson process the approximation of the true process by a torus process must be considered, and this seems impossible to quantify analytically.

(b) Border method. As before, choose E^* with $E^* \oplus b(\mathbf{0}, t) \subset E$, and apply the naive estimator to the m points lying within E^*. Thus

$$\hat{K}_1(t) = \frac{1}{\lambda m} \# [\text{pairs } (\mathbf{x}, \mathbf{y}) \text{ of } t\text{-close points with } \mathbf{x} \in E^*]$$

This is clearly at least approximately unbiased (since m is random).

(c) Isotropic correction (Ripley, 1976b). This correction is based on isotropy. Consider figure 3.3a, and consider just one pair $(\mathbf{x}, \mathbf{y})$. If $\partial b(\mathbf{x}, d(\mathbf{x}, \mathbf{y}))$ is not wholly contained within E, there could be unobserved points at distance $d(\mathbf{x}, \mathbf{y})$ from $\mathbf{x}$. It is this that causes the underestimation in $\hat{K}_0(t)$, and to compensate we count the pair $(\mathbf{x}, \mathbf{y})$ more than once, in fact $k(\mathbf{x}, \mathbf{y})$ times where

$$\frac{1}{k(\mathbf{x}, \mathbf{y})} = \frac{|\partial b(\mathbf{x}, d(\mathbf{x}, \mathbf{y})) \cap E|}{|\partial b(\mathbf{x}, d(\mathbf{x}, \mathbf{y}))|}$$

is the proportion of the perimeter of the circle which is within E. Note that in most cases $k(\mathbf{x}, \mathbf{y}) = 1$ since $\mathbf{x}$ will not be within distance t of E^c, at least for small t. Define

$$\hat{K}_2(t) = \frac{1}{\lambda^2 a} \Sigma k(\mathbf{x}, \mathbf{y}) 1[0 < d(\mathbf{x}, \mathbf{y}) \leqslant t]$$

Then $\hat{K}_2$ is unbiased provided E is convex and $t \leqslant t_0$, the circumradius of E. For, from (1),

$$\begin{aligned} \mathrm{E}\, \lambda^2 a K_2(t) &= \int k(\mathbf{x}, \mathbf{y}) 1[0 < d(\mathbf{x}, \mathbf{y}) \leqslant t] \mathrm{d}\mu_2(\mathbf{x}, \mathbf{y}) \\ &= \lambda^2 \int_0^\infty \left\{ \int k(\mathbf{x}, \mathbf{y}) 1[0 < d(\mathbf{x}, \mathbf{y}) \leqslant t] \mathrm{d}\nu_s(\mathbf{x}, \mathbf{y}) \right\} \mathrm{d}K(s) \\ &= \lambda^2 \int_0^t a \mathrm{d}K(s) = \lambda^2 a K(t) \quad \text{for } t \leqslant t_0 \end{aligned}$$

The effect of the correction factor k is to rescale the σ_s term in ν_s. This is possible if $E \cap \partial b(\mathbf{x}, s)$ has positive length for each $\mathbf{x} \in E$. For a convex set this will hold for $s \leqslant t_0$.

Unbiasedness up to t_0 will usually suffice; for a disc it is the radius and for a rectangle half the length of a diagonal. Nevertheless, Ohser (1983)

Figure 3.3 Calculation of edge corrections within a window E. (a) *Isotropic correction.* Here $1/k$ (x, y) is the proportion of the circle circumference which is shown as a solid line.. (b) *Translational correction.* The correction factor is the ratio of the area of E to the cross-hatched area.

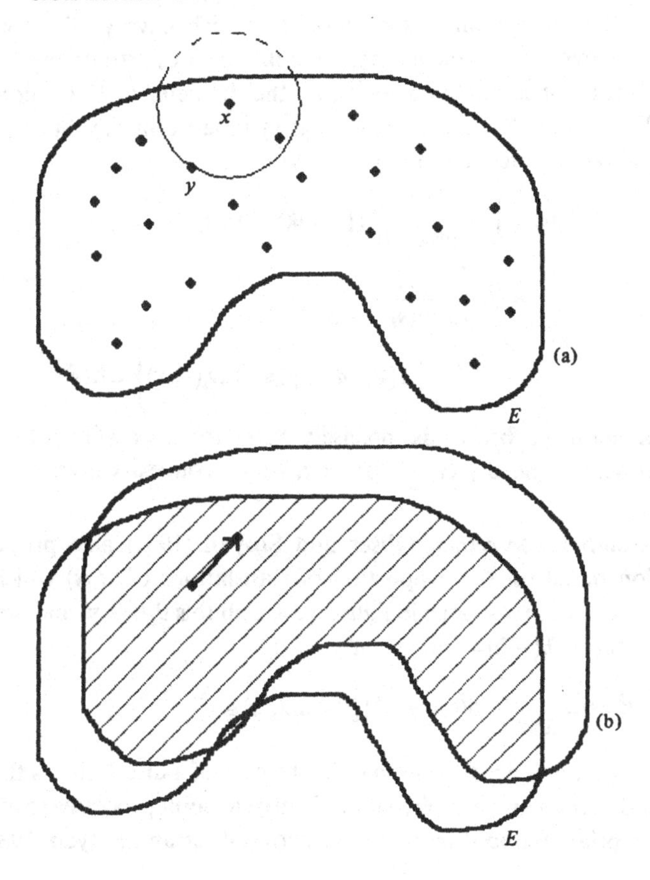

extended the range of unbiasedness by redefining

$$\hat{K}_2'(t) = \frac{1}{\lambda^2} \sum \frac{k(\mathbf{x}, \mathbf{y}) 1[0 < d(\mathbf{x}, \mathbf{y}) \leqslant t]}{v\{\mathbf{x} \mid \partial b(\mathbf{x}, d(\mathbf{x}, \mathbf{y})) \cap E \neq \varnothing\}}$$

This rescales the v_s term to a provided $v\{\mathbf{x} \mid \partial b(\mathbf{x}, s) \cap E \neq \varnothing\} > 0$, and so $\hat{K}_2'(t)$ is unbiased for $t \leqslant t_1$, the diameter of E, provided E is convex. Note that K_2' and K_2 agree for $t \leqslant t_0$.

(d) Rigid motion correction. This appears to be due to Miles (1974) (his 'minus sampling') for a disc, and to Ohser and Stoyan (1981) for general windows E. Again the pair (x, y) is counted more than once, in fact

$a/\gamma_E(d(\mathbf{x}, \mathbf{y}))$ times. Thus

$$\hat{K}_3(t) = \frac{1}{\lambda^2 a} \sum \frac{a}{\gamma_E(d(\mathbf{x}, \mathbf{y}))} 1[0 < d(\mathbf{x}, \mathbf{y}) \leq t]$$

The effect here is consider a rigid motion of $(\mathbf{x}, \mathbf{y})$ which keeps $\mathbf{x} \in E$, and to calculate the proportion of such motions which have $\mathbf{y} \in E$ (proportion being measured by the kinematic measure). This proportion is $\gamma_E(d(\mathbf{x}, \mathbf{y}))/a$.

This estimator is unbiased for $t \leq t_1$, the diameter of E, for convex E. Note that, unlike $\hat{K}_2$, the correction is symmetric in $\mathbf{x}, \mathbf{y}$ but is always nontrivial (i.e. greater than one). We have

$$\begin{aligned} \mathrm{E}\,\lambda^2 \hat{K}_3(t) &= \int \frac{1}{\gamma_E(d(\mathbf{x}, \mathbf{y}))} 1[0 < d(\mathbf{x}, \mathbf{y}) \leq t]\, \mathrm{d}\mu_2(\mathbf{x}, \mathbf{y}) \\ &= \lambda^2 \int_0^\infty \frac{1}{\gamma_E(s)} \\ &\quad \times \left\{ \int_{E \times E} 1[0 < d(\mathbf{x}, \mathbf{y}) \leq t]\, \mathrm{d}\nu_s(\mathbf{x}, \mathbf{y}) \right\} \mathrm{d}K(s) \end{aligned}$$

and the term in braces is precisely $\gamma_E(s)$ for $s < t$. Thus to ensure unbiasedness we need $\gamma_E(s) > 0$ for $s \leq t$, which convexity gives for $t \leq t_1$.

(e) Translation correction. Ohser and Stoyan (1981) also proposed a correction based on the proportion of translations of $(\mathbf{x}, \mathbf{y})$ which have both $\mathbf{x}$ and $\mathbf{y}$ in E, illustrated in figure 3.3b. Call this $\theta_E(\mathbf{x} - \mathbf{y})$ and note that it is symmetric, $\theta(-\mathbf{h}) = \theta(\mathbf{h})$. Then

$$\hat{K}_4(t) = \frac{1}{\lambda^2 a} \sum \theta(\mathbf{x} - \mathbf{y})^{-1} 1[0 < d(\mathbf{x}, \mathbf{y}) \leq t]$$

which is again unbiased. Note that for the case of a disc E this is the same as (d), and that it can be applied for anisotropic point processes to estimate the appropriate reduced moment measure (Ohser and Stoyan, 1981).

(f) Grouping. Rather than apply a correction to every distance which occurs, we could average the correction over a small group of distances and so hope to reduce the work. The approach of Vere-Jones (1978) is to group the rigid motion approach. Glass and Tobler (1971) applied an edge correction which essentially averages the isotropic correction factor for the pairs with interpoint distances in a given group. This is midway between the isotropic and rigid motion corrections, without the exactness of either!

Some general remarks apply to all these corrections. None of them can work well for large t. For the border method E^* and hence m will be small.

For the correction methods the correction factors could be large. For example, with a rectangle $k(\mathbf{x}, \mathbf{y}) \leqslant 4$, but for $t_0 < t < t_1$, K'_2 can have much larger correction factors when its denominator is much less than a. The more we average over, the more stable the correction factors, so rigid motion correction might be expected to give the least variable answers. Conversely, the less averaging we do, the more appropriate the correction will be to the particular pattern we observe.

It is not easy to predict the best balance between these two opposing effects. Analogous examples suggest that the answer might depend on the dimension. In considering covariances of lattice processes on $\mathbb{Z}^d$, it is normal not to make a translation-based edge correction (divide by $n - |\text{lag}|$ not n) for $d = 1$ but essential to do so for $d \geqslant 3$. For $d = 2$ an intermediate correction based on tapering may be best (Dahlhaus and Künsch, 1987).

3.3 ASYMPTOTIC VARIANCES FOR EDGE-CORRECTED ESTIMATES

In this section expectations and variances refer either to a Poisson process (suffix λ) or binomial process (suffix n) on $\mathbb{R}^2$. They are approximate for both small $t/\sqrt{a}$ and large λa or n. To be specific we will illustrate the case $E = b(\mathbf{0}, R)$, λ fixed, $R \to \infty$.

Some of these results are related to those in Ripley (1984a). That paper worked via U-statistics only for $E = b(\mathbf{0}, R)$, and considered estimators of $\lambda^2 K(t)$. Estimating $K(t)$ provides some critical differences.

3.3.1 Naive estimator

We have

$$\hat{K}_0(t) = aT/n^2$$

where

$$T = \Sigma\, 1[0 < d(\mathbf{x}, \mathbf{y}) \leqslant t]$$

summed over ordered pairs of observed points. Then from (9) and (11),

$$\mathrm{E}_n \hat{K}_0(t) = \frac{a}{n^2} \frac{n(n-1)S}{a^2} \tag{12}$$

$$\mathrm{var}_n \hat{K}_0(t) = \frac{2(n-1)}{n^3} \left[\frac{2(n-2)S_1}{a} - \frac{(2n-3)S^2}{a^2} + S_2 \right] \tag{13}$$

Now

$$S_2 = S = \int_0^t \gamma_E(s) 2\pi s \mathrm{d}s \approx a\pi t^2 - 2ut^3/3$$

as before. We seek a more accurate formula for S_1, based on approximating the boundary by a straight line. Now

$$S_1=\int_E\left\{\int_0^t|\partial b(\mathbf{x},s)\cap E|\,\mathrm{d}s\right\}^2\mathrm{d}\mathbf{x}$$

and $|\partial b(\mathbf{x},s)\cap E|=2\pi s$ for $s\leqslant t$ and $\mathbf{x}\in E^*=\{\mathbf{x}\,|\,b(\mathbf{x},t)\subset E\}$.

Thus

$$S_1=(\pi t^2)^2\nu(E^*)+\int_{E\setminus E^*}\{\ldots\}^2\,\mathrm{d}\mathbf{x}$$

and if we assume that the boundary is straight, $\nu(E^*)\approx a-ut$. Thus

$$S_1\approx(\pi t^2)^2\cdot(a-ut)+ut^5\int_0^1 f(h)^2\,\mathrm{d}h$$

where

$$f(h)=\nu(b(\mathbf{x}_h,1)\cap E)$$

for a point $\mathbf{x}_h$ distance h inside E. (This expression comes from rescaling the problem with a straight boundary so $t=1$.) We can evaluate f by

$$f(h)=\int_0^1\zeta(h,s)s\,\mathrm{d}s$$

$$\zeta(h,s)=\begin{cases}2\pi & s<h\\ 2(\pi-\cos^{-1}(h/s)) & s\geqslant h\end{cases}$$

Note that $0\leqslant f\leqslant\pi$ and $\int_0^1 f(h)\mathrm{d}h=\pi-\frac{2}{3}$. Thus

$$S_1\approx a\pi^2t^4+ut^5\left(\textstyle\int f^2-\pi^2\right)$$

and, from (13),

$$\begin{aligned}\mathrm{var}_n\hat{K}_0(t)&\approx\frac{2(n-1)}{n^3}\left[2(n-2)\left(\pi^2t^4+\frac{ut^5}{a}\left(\textstyle\int f^2-\pi^2\right)\right)\right.\\&\qquad\left.-(2n-3)\left(\pi t^2-\frac{2ut^3}{3a}\right)^2+a\pi t^2-\frac{2ut^3}{3}\right]\\&\approx\frac{2(n-1)}{n^3}\left[a\pi t^2-\frac{2ut^3}{3}+2n\left(\frac{4\pi}{3}-\pi^2+\int f^2\right)\frac{ut^5}{a}\right]\\&=\frac{2(n-1)}{n^3}\left[a\pi t^2-\frac{2ut^3}{3}+2n\int(\pi-f)^2\,\frac{ut^5}{a}\right]\\&\approx\frac{2(n-1)a^2}{n^3}\left[\frac{\pi t^2}{a}-\frac{2ut^3}{3a^2}+1.34\frac{nut^5}{a^3}\right]\end{aligned}$$

by numerical integration.

Thus for a binomial process

$$\left.\begin{aligned} \mathrm{E}_n\hat{K}_0(t) &\approx \left(1-\frac{1}{n}\right)\left(\pi t^2-\frac{2ut^3}{3a}\right) \\ \mathrm{var}_n\hat{K}_0(t) &\approx 2\left(\frac{a}{n}\right)^2\left[\frac{\pi t^2}{a}-\frac{2at^3}{3a^2}+1.34n\frac{ut^5}{a^3}\right] \end{aligned}\right\} \qquad \textbf{(14a)}$$

If we use the same form of $\hat{K}_0(t)$ but take expectations over n for a Poisson process we find

$$\left.\begin{aligned} \mathrm{E}_\lambda\hat{K}_0(t) &\approx \left(1-\frac{1}{\lambda}\right)\left(\pi t^2-\frac{2ut^3}{3a}\right) \\ \mathrm{var}_\lambda\hat{K}_0(t) &\approx \frac{2}{\lambda^2}\left[\frac{\pi t^2}{a}-\frac{2ut^3}{3a^2}+1.34\lambda\frac{ut^5}{a^2}\right] \end{aligned}\right\} \qquad \textbf{(14b)}$$

We can obtain a less biased estimator and essentially the same variance by dividing by $n(n-1)$ rather than by n^2. Note that we have apparently evaluated $\mathrm{E}_\lambda(n^{-1})$ and $\mathrm{E}_\lambda(n^{-2})$. This is not so, since $\hat{K}_0(t)=0$ if $n=0$ or 1. Table 3.1 shows that the approximations used here are reasonable.

Table 3.1

λa	$\lambda a\mathrm{E}[(1/n)1(n\geqslant 2)]$	$(\lambda a)^2\mathrm{E}[(n-1)/n^3]$	$(\lambda a)^3\mathrm{E}[(n-1)(n-2)/n^3]$
10	1.13	1.24	0.73
20	1.06	1.11	0.88
50	1.02	1.04	0.96

3.3.2 Rigid motion correction

The expressions for this estimator turn out to be very similar to those for the naive estimator, so it is most convenient to treat it next. We have

$$\hat{K}_3(t)=\frac{a}{n^2}\,\Sigma\phi(\mathbf{x},\mathbf{y})$$

where

$$\phi(\mathbf{x},\mathbf{y})=a1[0<d(\mathbf{x},\mathbf{y})\leqslant t]/\gamma_E(d(\mathbf{x},\mathbf{y}))$$

and

$$\gamma_E(s)\approx a-us$$

Now

$$S=\int_{E^2}\phi(\mathbf{x},\mathbf{y})\mathrm{d}\mathbf{x}\,\mathrm{d}\mathbf{y}=a\int_0^t\gamma_E(s)^{-1}\gamma_E(s)2\pi s\,\mathrm{d}s=a\pi t^2$$

so

$$\mathrm{E}_n\hat{K}_3(t)=\left(1-\frac{1}{n}\right)\pi t^2, \qquad \mathrm{E}_\lambda\hat{K}_3(t)\approx\left(1-\frac{1}{\lambda}\right)\pi t^2 \tag{15}$$

Also

$$S_2=\int_{E^2}\phi(\mathbf{x},\mathbf{y})^2\,\mathrm{d}\mathbf{x}\,\mathrm{d}\mathbf{y}=2\pi a^2\int_0^t s/\gamma_E(s)\,\mathrm{d}s \tag{16a}$$

$$\approx -2\pi^2a^2\left[\frac{t}{u}+\frac{\pi}{u^2}\ln\left(1-\frac{ut}{\pi a}\right)\right] \tag{16b}$$

$$\approx a\pi t^2+2ut^3/3 \tag{16c}$$

and

$$\mathrm{S}_1=\int_{E^*}\left\{a\int_0^t 2\pi s/\gamma_E(s)\right\}^2\mathrm{d}\mathbf{x} + \int_{E\setminus E^*}\left\{a\int_0^t\frac{|\partial b(\mathbf{x},s)\cap E|}{\gamma_E(s)}\mathrm{d}s\right\}^2\mathrm{d}\mathbf{x} \tag{17a}$$

$$\approx\left(\frac{S_2}{a}\right)^2\nu(E^*)+\int_{E\setminus E^*}\left\{\int_0^t|\partial b(\mathbf{x},s)\cap E|\right\}^2\mathrm{d}\mathbf{x}$$

since we need keep only the leading term of the second integral. This gives

$$S_1\approx(a-ut)\left(\pi t^2+\frac{2ut^3}{3a}\right)^2+ut^5\int f^2$$

$$=a\pi^2t^4+ut^5\left(\frac{4\pi}{3}-\pi^2+\int f^2\right)$$

$$\approx a\pi^2t^4+ut^5\int(\pi-f)^2$$

$$\approx a\pi^2t^4+0.67ut^5 \tag{17b}$$

From (13) we find

$$\mathrm{var}_n\,\hat{K}_3(t)\approx\frac{2(n-1)}{n^3}\left[\frac{2(n-2)}{a}(a\pi^2t^4+0.67ut^5) -(2n-3)\pi^2t^4+2a\pi t^2+\frac{4ut^3}{3}\right]$$

$$\approx\frac{2(n-1)}{n^3}\left[a\pi t^2+\frac{2ut^3}{3}+1.34(n-2)\frac{ut^5}{a}\right]$$

$$\approx 2\left(\frac{a}{n}\right)^2\left[\frac{\pi t^2}{a}+\frac{2ut^3}{3a^2}+1.34\left(\frac{n}{a}\right)\frac{ut^5}{a^2}\right] \tag{18a}$$

and for a Poisson process

$$\mathrm{var}_\lambda\,\hat{K}_3(t)\approx\frac{2}{\lambda^2}\left[\frac{\pi t^2}{a}+\frac{2ut^3}{3a^2}+1.34\lambda\frac{ut^5}{a^2}\right] \tag{18b}$$

Table 3.2 *Tests of the approximations for S_1 and S_2. The test windows are a disc of unit radius, the unit square and rectangles of aspect ratios 1:2 and 1:5 with unit area.*

		S_1				S_2		
		(17a)	π^2t^4/a	(17b)		(16a)	(16b)	(16c)
Circle								
$t=0.05$	$\times 10^{-5}$	1.977	1.963	1.977	$\times 10^{-5}$	2.554	2.554	2.553
0.10	$\times 10^{-4}$	3.184	3.142	3.184	$\times 10^{-2}$	1.044	1.044	1.042
0.25	$\times 10^{-2}$	1.27	1.23	1.27	$\times 10^{-2}$	7.00	7.00	6.91
0.50		0.208	0.196	0.210		0.319	0.320	0.303
Square								
0.05	$\times 10^{-5}$	6.256	6.168	6.253	$\times 10^{-3}$	8.20	8.20	8.18
0.10	$\times 10^{-4}$	10.14	9.87	10.14	$\times 10^{-2}$	3.43	3.44	3.41
0.25	$\times 10^{-2}$	4.11	3.86	4.12		0.248	0.251	0.238
Rect. 1:2								
0.05	$\times 10^{-5}$	6.256	6.168	6.257	$\times 10^{-3}$	8.22	8.23	8.21
0.10	$\times 10^{-4}$	10.15	9.87	10.15	$\times 10^{-2}$	3.45	3.46	3.42
0.25	$\times 10^{-2}$	4.12	3.86	4.14		0.252	0.256	0.241
Rect. 1:5								
0.05	$\times 10^{-5}$	6.277	6.168	6.280	$\times 10^{-3}$	8.33	8.33	8.30
0.10	$\times 10^{-4}$	10.21	9.87	10.23	$\times 10^{-2}$	3.55	3.55	3.50

Table 3.2 provides some tests of the approximations used in finding S_1 and S_2. These show that expressions (18) are rather accurate approximations.

3.3.3 Isotropic correction

This differs from the previous two cases in having different terms for $(\mathbf{x}, \mathbf{y})$ and $(\mathbf{y}, \mathbf{x})$. Thus we use

$$\hat{K}_2(t)=\frac{a}{n^2}\sum_{\mathbf{x},\mathbf{y}} 1(0<d(\mathbf{x},\mathbf{y})\leqslant t)\tfrac{1}{2}\{k(\mathbf{x},\mathbf{y})+k(\mathbf{y},\mathbf{x})\}$$

Let

$$\psi(\mathbf{x},\mathbf{y})=1(0<d(\mathbf{x},\mathbf{y})\leqslant t)k(\mathbf{x},\mathbf{y})$$

so

$$\phi(\mathbf{x},\mathbf{y})=\tfrac{1}{2}\{\psi(\mathbf{x},\mathbf{y})+\psi(\mathbf{y},\mathbf{x})\}$$

Now

$$\begin{aligned}S&=\int_{E^2}\phi(\mathbf{x},\mathbf{y})\mathrm{d}\mathbf{x}\,\mathrm{d}\mathbf{y}=\int_{E^2}\psi(\mathbf{x},\mathbf{y})\mathrm{d}\mathbf{x}\,\mathrm{d}\mathbf{y}\\&=\int_E\int_0^t\left\{\int_0^{2\pi}1(\mathbf{y}\in E)\mathrm{d}\theta\,k(\mathbf{x},\mathbf{y})\right\}2\pi s\,\mathrm{d}s\,\mathrm{d}\mathbf{x}=a\pi t^2\end{aligned}$$

where $\mathbf{y}=\mathbf{x}+t(\cos\theta, \sin\theta)$. Now

$$S_2=\int \phi(\mathbf{x}, \mathbf{y})^2\, d\mathbf{x}\, d\mathbf{y}$$

$$=\tfrac{1}{2}\int \psi(\mathbf{x}, \mathbf{y})^2\, d\mathbf{x}\, d\mathbf{y}+\tfrac{1}{2}\int \psi(\mathbf{x}, \mathbf{y})\psi(\mathbf{y}, \mathbf{x})\, d\mathbf{x}\, d\mathbf{y}$$

and since $\psi(\mathbf{x}, \mathbf{y})$ is 1 unless $\mathbf{y}$ is within distance $2t$ of ∂E,

$$S_2\approx \pi t^2(a-2ut)$$

$$+ut\,.\,\pi t^2\left[1+\int_0^1 \chi(h, \mathbf{y})^2\, d\mathbf{y}\, dh+\int_0^2 \chi(h, \mathbf{y})\chi(\mathbf{y}, h)\, d\mathbf{y}\, dh\right]$$

$$\approx 2a\pi t^2\left(1+0.305\,\frac{ut}{a}\right)$$

using $\chi(h, \mathbf{y})=\psi(\mathbf{x}_h, \mathbf{y})$ for $\mathbf{x}_h$ distance h from the boundary. Similarly,

$$S_1=\int\left\{\int \phi(\mathbf{x}, \mathbf{y})d\mathbf{y}\right\}^2 d\mathbf{x}=\int\left\{\int \phi(\mathbf{x}, \mathbf{y})d\mathbf{y}-\pi t^2\right\}^2 d\mathbf{x}$$

$$+2a\pi t^2\iint \phi(\mathbf{x}, \mathbf{y})d\mathbf{x}\, d\mathbf{y}-a(\pi t^2)^2$$

$$=a\pi^2t^4+\int\left\{\int \phi(\mathbf{x}, \mathbf{y})d\mathbf{y}-\pi t^2\right\} d\mathbf{x}$$

and $\{\ldots\}$ is non-zero only when $\mathbf{x}$ is within distance $2t$ of ∂E, so

$$S_1\approx a\pi^2t^4+ut(\pi t^2)^2\int_0^2\left\{\int \chi(h, \mathbf{y})d\mathbf{y}-1\right\}^2 dh$$

$$\approx a\pi^2t^4(1+0.0066ut/a)$$

Collecting the pieces gives

$$\operatorname{var}_n \hat{K}_2(t)\approx\frac{2(n-1)}{n^3}\left[2(n-2)\pi^2t^4\left(1+0.0066\,\frac{ut}{a}\right)\right.$$

$$\left.-(2n-3)\pi^2t^4+2a\pi t^2\left(1+0.305\,\frac{ut}{a}\right)\right]$$

$$\approx 2\left(\frac{a}{n}\right)^2\left[\frac{\pi t^2}{a}+0.96\,\frac{ut^3}{a^2}+0.13\left(\frac{n}{a}\right)\frac{ut^5}{a^2}\right] \qquad \textbf{(19a)}$$

and

$$\operatorname{var}_\lambda \hat{K}_2(t)\approx\frac{2}{\lambda^2}\left[\frac{\pi t^2}{a}+\frac{0.96ut^3}{a^2}+0.13\lambda\,\frac{ut^5}{a^2}\right] \qquad \textbf{(19b)}$$

3.3.4 Border correction

Suppose we observe m points in E^* and $n\geqslant m$ points in E. Then a natural form for the border-corrected estimator is

$$\hat{K}_1(t)=\frac{a}{mn}\,\#\,(t\text{-close pairs with } \mathbf{x}\in E^*,\ \mathbf{y}\in E)$$

Ripley (1984a) computed $\operatorname{var}\{mn\hat{K}_1(t)\}$ by U-statistics methods for a disc. Since m is random and correlated with $\hat{K}_1(t)$, this is of little help here. We take a more direct approach. Let $r=n-m$. Then

$$\hat{K}_1(t)=\frac{a}{mn}\left[\sum_{1\leqslant i<j\leqslant m} 2\phi(\mathbf{X}_i,\mathbf{X}_j)+\sum_{i=1}^{m}\sum_{j=1}^{r}\phi(\mathbf{X}_i,\mathbf{Y}_i)\right]$$

where $(\mathbf{X}_1,\ldots,\mathbf{X}_m,\mathbf{Y}_1,\ldots,\mathbf{Y}_r)$ are independent, the $\mathbf{X}$'s uniform on E^* and the $\mathbf{Y}$'s uniform on $E\backslash E^*$. Here $\phi(\mathbf{x},\mathbf{y})=1[0<d(\mathbf{x},\mathbf{y})\leqslant t]$. First fix m (and hence r). Then

$$\mathrm{E}[\hat{K}_1(t)|m]=\frac{a}{mn}\left[\frac{m(m-1)}{a^{*2}}\int_{E^*\times E^*}\phi+\frac{rm}{a^*(a-a^*)}\int_{E^*\times(E\backslash E^*)}\phi\right]$$

which under the straight boundary approximation gives

$$\begin{aligned}\mathrm{E}[\hat{K}_1(t)|m]&\approx\frac{a}{mn}\left[\frac{m(m-1)}{a^{*2}}\left(a^*\pi t^2-\frac{2u^*t}{3}\right)+\frac{rm}{uta^*}\frac{2}{3}u^*t^3\right]\\&\approx\frac{a}{a^*n}\left[(m-1)\left(\pi t^2-\frac{2ut^3}{3a}\right)+\frac{2rt^2}{3}\right]\end{aligned}$$

Now $r\sim$ binomial $(n, ut/a)$ so

$$\begin{aligned}\mathrm{E}_n\hat{K}_1(t)&\approx\frac{a}{a^*n}\left[\left(n-1-\frac{nut}{a}\right)\left(\pi t^2-\frac{2ut^3}{3a}\right)+\frac{2nut^3}{3a}\right]\\&\approx\frac{[n(1-ut/a)-1]}{na^*}\pi t^2+\mathrm{O}(t^4)\approx\left(1-\frac{1}{n}\right)\pi t^2+\mathrm{O}(t^4)\end{aligned}$$

Thus $\hat{K}_1(t)$ is approximately unbiased. To find $\operatorname{var}_n\hat{K}_1(t)$ we will compute $\operatorname{var}\mathrm{E}[\hat{K}_1(t)|m]$ and $\mathrm{E}\operatorname{var}[\hat{K}_1(t)|m]$. From the above

$$\operatorname{var}\mathrm{E}[\hat{K}_1(t)|m]\approx(\pi-\tfrac{2}{3})^2\cdot\frac{ut^5}{na}$$

For the conditional variance, let

$$A=2\sum_{i<j}\phi(\mathbf{X}_i,\mathbf{X}_j),\qquad B=\sum_{i,j}\phi(\mathbf{X}_i,\mathbf{Y}_j)$$

Then

$$\operatorname{var}(A)\approx 2m^2\left[\frac{\pi t^2}{a^*}-\frac{2ut^3}{3a^{*2}}+2\int(\pi-f)^2\,\frac{mut^5}{a^{*3}}\right]$$

from §3.3.1. Further

$$\begin{aligned}\operatorname{cov}(A,B)&=2rm(m-1)\operatorname{cov}[\phi(\mathbf{X}_1,\mathbf{X}_2),\phi(\mathbf{X}_1,\mathbf{Y}_1)]\\&\approx 2rm(m-1)\left[\frac{ut^5}{uta^{*2}}\int_0^1 f(h)\{\pi-f(h)\}\mathrm{d}h\right.\\&\qquad\left.-\left(\pi t^2-\frac{2ut^3}{a}\right)\left(\frac{2t^2}{3}\right)\Big/a^{*2}\right]\end{aligned}$$

$$\approx \frac{2rm^2t^4}{a^{*2}}\left[\pi\int f - \int f^2 - \tfrac{2}{3}\pi\right] = -\frac{2rm^2t^4}{a^{*2}}\int(\pi - f)^2$$

$$\text{var}(B) = rmE[\phi(\mathbf{X}_1, \mathbf{Y}_1)^2] + r(r-1)mE[\phi(\mathbf{X}_1, \mathbf{Y}_1)\phi(\mathbf{X}_1, \mathbf{Y}_2)]$$

$$+ rm(m-1)E[\phi(\mathbf{X}_1, \mathbf{Y}_1)\phi(\mathbf{X}_2, \mathbf{Y}_1)] - rm(n-2)[E\phi(\mathbf{X}, \mathbf{Y})]^2$$

$$\approx \frac{rm2t^2}{3a^*} + \frac{r^3mt^3}{ua^*}\int(\pi - f)^2 + \frac{rm^2t^4}{a^{*2}}\int(\pi - f)^2 - \frac{rmn4t^4}{9a^{*2}}$$

Thus

$$\text{E var}\,\frac{A}{m} \approx 2\left[\frac{\pi t^2}{a^*} - \frac{2ut^3}{3a^{*2}} + 1.34\,\frac{nut^5}{a^3}\right]$$

$$\text{E cov}\left(\frac{A}{m}, \frac{B}{m}\right) \approx -\frac{2nut^5}{a^3}\int(\pi - f)^2$$

$$\text{E var}\,\frac{B}{m} \approx \frac{2ut^3}{3a^2} + \frac{t^4}{a^2}\int(\pi - f)^2 + \frac{nut^5}{a^3}\int(\pi - f)^2 - \frac{4nut^5}{9a^3}$$

so

$$\text{E var}[\hat{K}_1(t)|m] \approx \frac{2a^2}{n^2}\left[\frac{\pi t^2}{a^*} - \frac{ut^3}{3a^2}\right.$$

$$\left. + \frac{nut^5}{a^3}\left\{1.34 + 2\cdot(-0.67) + \tfrac{1}{2}\int(\pi - f)^2 - \tfrac{2}{9}\right\}\right]$$

$$\approx \frac{2a^2}{n^2}\left[\frac{\pi t^2}{a} + (\pi - \tfrac{1}{3})\,\frac{ut^3}{a^2} + 0.11\,\frac{nut^5}{a^3}\right]$$

and

$$\text{var}_n\hat{K}_1(t) \approx 2\left(\frac{a}{n}\right)^2\left[\frac{\pi t^2}{a} + (\pi - \tfrac{1}{3})\,\frac{ut^3}{a^2} + 3.17\,\frac{nut^5}{a^3}\right] \qquad (20)$$

3.3.5 Discussion

All these expressions reduce to the form

$$\text{var}_\lambda\hat{K}_\alpha(t) \approx \frac{2}{\lambda^2}\left[\frac{\pi t^2}{a} + c_1\,\frac{ut^3}{a^2} + c_2\,\frac{\lambda ut^5}{a^2}\right]$$

The first term corresponds to a Poisson limit, the sum being made up of $\frac{1}{2}n(n-1)$ rare events of probability $\pi t^2/a^2$ each. We will see in §3.4 the corresponding limit theorem. The other two terms are due to edge effects and correlations between the rare events respectively, these correlations occurring at the edge to the level of accuracy pursued here. The relative importance of these terms depends on $\lambda\pi t^2$, the expected number of points

Table 3.3. *Comparison of approximate variances of edge-corrected estimators of K(t) for a binomial process of n points in a disc of radius R. Entries are factor times Poisson approximation.*

t/R	Border	Ripley	Ohser–Stoyan
n=20			
0.10	1.185	1.061	1.045
0.25	1.555	1.157	1.149
n=100			
0.05	1.097	1.030	1.024
0.10	1.243	1.063	1.070
0.25	2.450	1.194	1.530
n=500			
0.05	1.129	1.032	1.038
0.10	1.500	1.074	1.178
0.25	6.465	1.359	3.228

t-close to an existing point. For a disc $E=b(\mathbf{0}, R)$ we have

$$\operatorname{var}_\lambda \hat{K}_\alpha(t) \approx \frac{2}{\lambda^2}\left[\pi\left(\frac{t}{R}\right)^2 + \frac{2c_1}{\pi}\left(\frac{t}{R}\right)^3 + \frac{2c_2}{\pi^2}\cdot\lambda\pi t^2\cdot\left(\frac{t}{R}\right)^3\right]$$

Table 3.3 shows the effects of the differences in c_1 and c_2 in various cases. This shows the relative inefficiency of the border correction, and that the isotropic correction is best when $\lambda\pi t^2$ is large, otherwise it and the Ohser–Stoyan correction are very similar.

The reader may wonder why we have given no separate results for the translation correction. Of course, in the case illustrated in table 3.3 it agrees with the rigid body correction. In general an analysis similar to that given here depends on much more than the area and perimeter of E, and the results do not appear enlightening.

The results given here differ from those of Ripley (1984a) both in generality and in working with estimators of $K(t)$ not $\kappa(t)=\lambda^2K(t)$. Further, the results in that paper are wrong as far as the Poisson process is concerned, although correct for a binomial process. They are in fact expressions for E $\operatorname{var}_n\hat{\kappa}(t)$ and in each case should be added.

$$\operatorname{var}_\lambda \mathrm{E}(\hat{\kappa}(t)\mid n) = 4\lambda^3\pi^2t^4/a$$

since

$$\hat{\kappa}(t) = \frac{n^2}{a^2}\hat{K}(t)$$

for $\hat{K}_2$ and $\hat{K}_3$ at least. The main differences are for the border-corrected estimator. Then

$$\hat{\kappa}(t) = \frac{\#(t\text{-close pairs with } \mathbf{x} \in E^*)}{\nu(E^*)} = \frac{mn}{aa^*}\hat{K}_1(t)$$

and the full expression becomes

$$\operatorname{var}_\lambda \hat{\kappa}(t) = 2\lambda^2\left[\frac{\pi t^2}{a} + (\pi - \tfrac{1}{3})\frac{ut^3}{a^2} + 2\lambda\left(\frac{2\pi^2 t^4}{a} + \frac{8.1ut^5}{a^2}\right)\right]$$

since the size of the term $\operatorname{var}\{\mathrm{E}[\hat{\kappa}(t) \mid m]\}$ differs from that for $\hat{K}_1(t)$.

3.4 LIMIT THEOREMS FOR INTERPOINT DISTANCES

Let $Y(t)$ denote the number of t-close pairs in E, so $\hat{K}_0(t) \propto Y(t)$. We have already seen that for a Poisson process

$$\mathrm{E}\,Y(t) \approx \tfrac{1}{2}\lambda^2 a\pi t^2$$

$$\operatorname{var} Y(t) \approx \mathrm{E}\,Y(t)$$

for t small compared to the size of E. Since $Y(t)$ is a count this suggests $Y(t)$ will be approximately Poisson distributed. We can deduce more accurate statements from the results of §3.3. We find

$$\mathrm{E}\,Y(t) = \tfrac{1}{2}\lambda^2 \int_0^t \gamma_E(s)\,\mathrm{d}s \approx \tfrac{1}{2}\lambda^2 a\pi t^2\left(1 - \frac{2ut}{3a}\right)$$

$$\operatorname{var} Y(t) = \lambda^3 S_1 + \tfrac{1}{2}\lambda^2 S_2$$

$$\approx \tfrac{1}{2}\lambda^2\left[\int_0^t \gamma_E(s)\,\mathrm{d}s + 2\lambda a\pi^2 t^4\right]$$

from (10), so

$$\frac{\operatorname{var} Y(t)}{\mathrm{E}\,Y(t)} \approx 1 + \frac{2\lambda a\pi^2 t^4}{a\pi t^2} = 1 + 2\lambda\pi t^2$$

For a *binomial* process we have

$$\frac{\operatorname{var} Y(t)}{\mathrm{E}\,Y(t)} \approx 1 + 1.34\frac{nut^3}{a^2} \tag{21}$$

from (14a). In the Poisson case the approximation is good provided the expected number of points within a disc of radius t, $\lambda\pi t^2$, is small.

A number of authors have proved Poisson limit theorems under these conditions. By the scale equivariance of the Poisson process we can vary t, λ and E together as $\lambda a \to \infty$ (so $n \to \infty$). Provided $\lambda^2 a t^2 \to$constant we obtain a Poisson limit for $Y(t)$. For a binomial process we need $n \to \infty$, $nt/\sqrt{a} \to$constant. Saunders and Funk (1977) proved slightly weaker versions of these results by showing the convergence of all moments to that of a Poisson distribution. More elegant proofs were given by Silverman and Brown (1978) and Brown and Silverman (1979). These authors all considered heterogeneous processes, and weak convergence of $Y(\ )$ on $[0, t_0]$ to a heterogeneous Poisson process.

Other asymptotic regimes are possible. Kester (1975) shows asymptotic normality of $Y(t)$ as $n \to \infty$ for a binomial process on a unit square provided

$$t_n \to 0, \quad nt_n \to \infty, \quad nt_n^3 \to 0$$

in contrast to the Poisson limit for $nt_n \to$constant. The condition $nt_n^3 \to 0$ is precisely that needed to ensure that the first term of (21) (and of 14a) predominates. Kester's proof is based on moments for a torus, followed by showing that edge effects on the square are asymptotically negligible. Silverman (1976) gives asymptotic normality for $n \to \infty$, t constant. Here edge effects do dominate. Different limits occur on sets without edges such as the sphere and the torus, at rate $1/n$ rather than $1/\sqrt{n}$ (Silverman, 1978), with non-normal limits.

Almost all these results can be deduced from earlier work of Eberl and Hafner (1971). This is based on systematic computation of moments and their convergence to those of a Poisson or normal distribution. The conditions of that paper are rather general and abstract, so some work is needed to show that they are satisfied under the conditions given here.

How do these results apply to $\hat{K}_i(t)$? In the Poisson limit the edge effects become negligible. Consider a disc $E = b(\mathbf{0}, R)$. Then for the Poisson limit if λ is fixed we need both $R \to \infty$ and $t \propto 1/R$, in which case edge effects are of order $(2\pi R t)/\pi R^2 = 2t/R = O(R^{-2})$ and so are negligible. We can see though that the Poisson approximation might work better for small but finite t if var $\hat{K}_i(t)$ is closer to the appropriate multiple of E $\hat{K}_i(t)$. We had, for a Poisson process,

$$\mathrm{E}_\lambda \hat{K}_i(t) \approx \pi t^2, \quad i = 1, 2, 3$$

$$\mathrm{var}_\lambda \hat{K}_i(t) \approx \frac{2}{\lambda^2 a}\left[\pi t^2 + \frac{c_1 u t^3}{a} + \frac{c_2 \lambda u t^5}{a}\right]$$

and

$$\frac{\mathrm{var}_\lambda \hat{K}(t)}{\mathrm{E}_\lambda \hat{K}_i(t)} \approx \frac{2}{\lambda^2 a}\left[1 + \frac{c_1 u t}{\pi a} + \frac{c_2 \lambda u t^3}{\pi a}\right]$$

We need both $c_1 ut/\pi a$ and $c_2 \lambda u t^3/a^2$ to be small, and those edge corrections with constants smaller than $\hat{K}_0(t)$ will probably give estimators more accurately proportional to a Poisson random variable. To see the size of this effect, consider n points within a disc of radius R. Then

$$\frac{\lambda u t^3}{a} \approx \frac{n \cdot 2\pi R t^3}{\pi^2 R^4} = n\frac{2}{\pi}\left(\frac{t}{R}\right)^3$$

and so for $t=0.2R$ the final term is less than 0.1 for $n \leqslant 19$ for the border correction but $n \leqslant 475$ for Ripley's correction. This has been exploited to provide a confidence band for $\hat{K}_2(t)$.

Since $\mathrm{E}\hat{K}_2(t)=\pi t^2$ for a Poisson process, it has become conventional to plot $\hat{K}_2(t)$ on a square-root scale. Let

$$\hat{L}(t)=\sqrt{[K_2(t)/\pi]}$$

so that $\hat{L}(t) \approx t$ for a Poisson (or binomial) process. Further, var $\hat{L}(t)$ will be approximately constant by the variance stabilizing properties of the square-root transformation, for small t at least. This suggests a confidence band of the sort

$$|\hat{L}(t)-t| \leqslant c$$

say (Ripley, 1979). We can re-express this as

$$\max(0, (t-c)^2) \leqslant \hat{K}_2(t) \leqslant \pi(c+t)^2 \qquad (22)$$

Since we have an approximate Poisson process for $\frac{1}{2}\lambda^2 a\hat{K}_2(t)$, we would expect $c \propto 1/\lambda\sqrt{a}=\sqrt{a}/n$. Strictly this result applies to an interval $t \in [0, t_0/\lambda\sqrt{a}]$ due to the scaling of the Poisson limit theorem. Rather fortuitously, $\hat{L}(t)$ is most variable for small t, due to the failure of the variance stabilization to be exact, and we can apply the same value of c for a wide range of t_0 with little change in

$$\mathrm{P}(\max_{t \leqslant t_0} |\hat{L}(t)-t| \leqslant c)$$

Simulations based on (22) show that $c=1.45\sqrt{a}/n$ gives a probability of about 95% provided nut_0^3/a^2 is small.

Limit theorems in this field have proven to be little help in setting bounds on approximations. Even the rate of convergence results to a Poisson limit of Brown and Silverman do not give explicit constants, despite attempts by the authors. This explains why much of this chapter has been devoted to careful evaluation of means and variances as the best available way to compare different estimators.

The role of conditioning

A perennial question when doing statistical work with point processes is whether we should condition on n, the number of points observed. For a Poisson process, n is the minimal sufficient statistic for λ, and is clearly informative. Where we are interested in interactions, n may be approximately ancilliary. For example, consider the Strauss process (discussed in more detail in chapter 4). This is defined by a Radon–Nikodym derivative relative to a Poisson process of

$$f(\mathbf{x})=a_0 b^{n(\mathbf{x})} c^{t(\mathbf{x})}$$

where $t(\mathbf{x})$ is the number of R-close pairs in the sample $\mathbf{x}$. This is defined on a bounded set E and hence is not stationary. We can still define an intensity $\lambda = \mathrm{E}[n(\mathbf{x})/a]$, as a function of both b and c. The minimal sufficient statistic is $(n(\mathbf{x}), t(\mathbf{x}))$. The exact marginal distribution of $n(\mathbf{x})$ is unknown. However, the sparse data approximation given above suggests that, for small R, the conditional distribution of $t(\mathbf{x})$ given n is approximately Poisson, with mean $n(n-1)\pi R^2/2$. We can view the Strauss process as obtained by rejection sampling of a Poisson process of intensity ba, the rejection probability being $c^{t(\mathbf{x})}$ for sample $\mathbf{x}$. Thus samples of size n are rejected with probability

$$\mathrm{E}[c^{t(\mathbf{x})} \mid n(\mathbf{x}) = n] \approx \exp[(c-1)n(n-1)\pi R^2/2a]$$

and

$$\mathrm{P}(n(\mathbf{x}) = n) \propto \frac{(ba)^n}{n!} \exp[(c-1)n(n-1)\pi R^2/2a] \tag{23}$$

The distribution of $n(\mathbf{x})$ does depend on c, but the dependence is quite weak in typical cases (figure 3.4). Thus we may feel justified in conditioning on n when considering interactions.

This is relevant to the variances of $\hat{K}(t)$ computed in this chapter. We defined $\hat{K}_0(t)$ in §3.2 involving λ, but since λ is usually unknown, our results are all with λ replaced by $\hat{\lambda} = n/a$. This applies equally to $\hat{K}_2$, $\hat{K}_3$,

Figure 3.4 Cumulative distribution functions of $n(\mathbf{X})$ for a Strauss process with $c = 0.0, 0.2, 0.4, 0.6, 0.8, 1.0$ from left to right. These are for $b = 50$, $R = 0.05$ within the unit square, computed from the approximation (23).

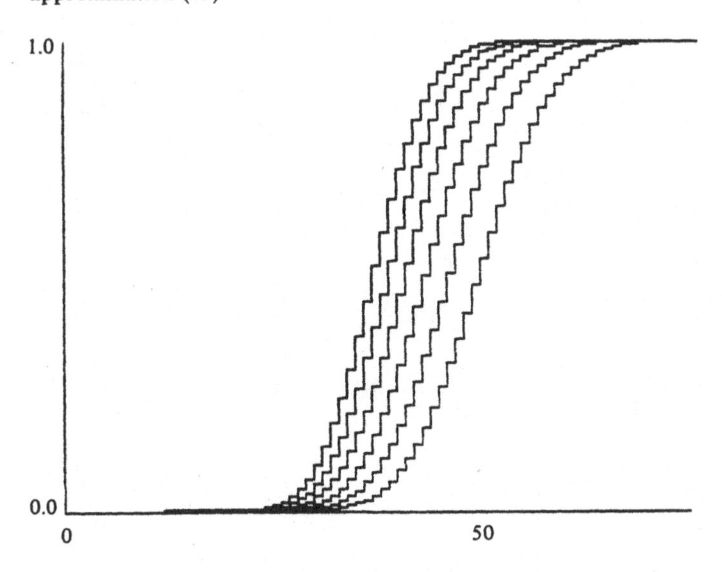

$\hat{K}_4$, all of which involved a factor λ^{-2}. Suppose we knew λ. What is the effect of using it rather than $\hat{\lambda}$? For definiteness consider $\hat{K}_2$. By using

$$\hat{K}_2(t) = \frac{1}{\lambda^2 a} \Sigma k(\mathbf{x}, \mathbf{y}) 1[0 < d(\mathbf{x}, \mathbf{y}) < t]$$

we gain exact unbiasedness. We know that for a Poisson process

$$\mathrm{var}_n \hat{K}_2(t) \approx \frac{2n^2}{\lambda^4 a^2} \left[\frac{\pi t^2}{a} + 0.96 \frac{ut^3}{a^2} + 0.13 \left(\frac{n}{a}\right) \frac{ut^5}{a^2} \right]$$

on rescaling (19a). However,

$$\mathrm{E}[\hat{K}_2(t) \mid n] = \frac{(n-1)}{\lambda a} \pi t^2, \quad n \geqslant 2$$

so

$$\mathrm{var}_\lambda \hat{K}_2(t) \approx \frac{(\pi t^2)^2}{\lambda a} + \frac{2}{\lambda^2} \left[\frac{\pi t^2}{a} + 0.96 \frac{ut^3}{a^2} + 0.13\lambda \frac{ut^5}{a^2} \right]$$

The additional term which arises from using λ rather than $\hat{\lambda}$ is $\frac{1}{2}\lambda\pi t^2$ times the Poisson limit approximation. This term is usually appreciable. The factor is half the number of t-close points to an arbitrary point, and will often be in the range 1–10. (In table 3.3 with n giving the expected number of points, the factor ranges from 0.31 to 49.) Thus using λ (known) rather than $\hat{\lambda}$ increases the variance of $\hat{K}_2(t)$ appreciably. Similar calculations apply to all the other estimators, and support conditioning on n even when λ is known.

4

Parameter estimation for Gibbsian point processes

Chapter 3 concentrated on the null hypothesis models of 'randomness', the binomial and Poisson processes. One of the advantages of techniques which produce a graphical summary such as $\hat{K}(\ )$ is that they will suggest (identify, in time series terminology) alternative models. We saw a few examples in chapter 1, together with the estimation problems of Poisson cluster processes. Here we concentrate on a family of models for interactions between points, the so-called Gibbsian models borrowed and adapted from statistical physics.

4.1 DEFINITIONS

Suppose we have a (stationary) point process on $\mathbb{R}^d$ which is observed within a window E of area a. We will see a random number N of points. A model must specify $P(N=n)$ and the p.d.f. $p_n(\mathbf{x}_1, \ldots, \mathbf{x}_n)$ conditional on $N=n$. We can always write

$$p_n(\mathbf{x}_1, \ldots, \mathbf{x}_n) = \frac{1}{Z} \exp[-\beta U(\mathbf{x}_1, \ldots, \mathbf{x}_n)] \tag{1}$$

where $U(\ )$ is the *energy* of that configuration of points and $\beta = 1/kT$ where T is the absolute temperature and k is the Boltzmann's constant.

We will need two extensions to (1). First consider variations in n, the grand canonical ensemble. Denote $\mathbf{x} = \{\mathbf{x}_1, \ldots, \mathbf{x}_n\}$, the set of observed points. Then it is most convenient to give the p.d.f. of $\mathbf{x}$ relative to a Poisson process of rate μ on E. That is

$$P(N=n) = e^{-\mu a}(\mu a)^n/n!$$

$$p_n(\mathbf{x}) = a^{-n} \text{ on } E^n$$

Then let f be the p.d.f. (Radon–Nikodym derivative) of the process defined by (1).

We find

$$f(\mathbf{x})=\frac{P(N=n)}{(e^{-\mu a}\mu^n/n!)}\exp[-U_n(\mathbf{x})] \qquad \text{if } \#\mathbf{x}=n$$

and so

$$f(\mathbf{x})=\exp[-U(\mathbf{x})] \tag{2}$$

for a suitable function U defined on $\mathscr{X}$, the set of finite subsets of E. (Details of this construction are given by Ripley and Kelly, 1977.)

The simplest forms of (2) are pairwise interaction processes, with

$$U(\mathbf{x})=a_0+\sum_i \psi(\mathbf{x}_i)+\sum_{i<j} \phi(\mathbf{x}_i, \mathbf{x}_j)$$

so the 'energy' is computed from points one or two at a time. If the process is to be stationary we will need $\psi(\ )\equiv a_1$, $\phi(\mathbf{x}, \mathbf{y})=\phi(d(\mathbf{x}, \mathbf{y}))$. We then find

$$f(\mathbf{x})=a_0 b^n \prod_{i<j} h(d(\mathbf{x}_i, \mathbf{x}_j)) \tag{3}$$

which specializes for $N=n$ to the pairwise interaction process considered in chapter 1. These processes have a *conditional intensity* $\lambda^*(\xi; \mathbf{x})$ of a point at ξ given $\mathbf{x}$ on $E\backslash\{\xi\}$ as

$$\lambda^*(\xi; \mathbf{x})=f(\mathbf{x}\cup\{\xi\})/f(\mathbf{x})=b \prod_{\eta\in\mathbf{x}} h(d(\xi, \eta)) \tag{4}$$

For $\mu=1$ this can be interpreted as the p.d.f. of a point at ξ conditional on $\mathbf{x}$ in $E\backslash\{\xi\}$.

Note that (3) will not always define a valid p.d.f. on $\mathscr{X}$. Some integrability conditions are needed to ensure that a_0 can be chosen appropriately. These amount to

$$\sum_{n=2}^{\infty} \frac{1}{n!} \int_{E^n} \prod_{i<j} h(d(\mathbf{x}_i, \mathbf{x}_j)) \prod d\mathbf{x}_i < \infty$$

To estimate parameters in h by maximum likelihood we can avoid a lot of these subtleties by conditioning on $N=n$, to obtain p.d.f.

$$p_n(\mathbf{x}_1, \ldots, \mathbf{x}_n)=\frac{1}{Z_\theta} \prod_{i<j} h_\theta(d(\mathbf{x}_i, \mathbf{x}_j)) \tag{5}$$

where

$$Z_\theta=\int_{E^n} \prod_{i<j} h_\theta(d(\mathbf{x}_i, \mathbf{x}_j)) \prod d\mathbf{x}_i$$

The problem here is that Z_θ is unknown and must be estimated or approximated.

A quite flexible family of functions h is

$$h(r)=\begin{cases} c_1 & 0<r\leqslant R_1 \\ c_2 & R_1<r\leqslant R_2 \\ \vdots & \\ c_m & R_{m-1}<r\leqslant R_m \\ 1 & r>R_m \end{cases}$$

sometimes called the multiscale family (Penttinen, 1984). As special cases this has:

(i) $m=1$, the so-called Strauss process defined by Kelly and Ripley (1976) following earlier incorrect work by Strauss (1975). This is characterized by

$$\lambda^*(\xi;\,\mathbf{x})=bc^{\#(\text{points of } \mathbf{x} \text{ within distance } R \text{ of } \xi)}$$

(ii) $m=1$, $c=0$, a hard-core inhibition process. Here realizations of a Poisson process on E are rejected completely if any pair of points is closer than R. In this case the maximum likelihood estimator of R is quite clearly $d_{\min}$, the minimum interpoint distance (Ripley and Silverman, 1978).

(iii) $m=2$, $c_1=0$; Saunders, Kryscio and Funk (1982).

Apart from these special cases, the general form is useful as an approximation to a general interaction function h, used in identification.

Thus far we have only considered processes defined on E. Such processes cannot be stationary, and (3) and (5) implicitly assume that there are no points outside E. The edge effect problem recurs! What we need to do is to define a p.d.f. f on $\mathscr{X}$ conditional on the points outside E, and check that this family of p.d.f.s for all E is consistent with a probability measure on the set of locally finite subsets of $\mathbb{R}^d$. This is a highly technical process with no simple solution. There may be no, one or many probability measures consistent with the specified conditional distributions. The case of many provides a mathematical model of *phase transition*, the coexistence of two phases at the same temperature. If *no* process exists this usually implies that an infinite set of points under energy U would collapse to accumulation points if released. Fortunately, we will not usually need to consider such behaviour, since in almost all examples $h(r)=1$ for $r>R$. Then all we will need is the conditional distribution of points in E given the points in $[E\oplus b(\mathbf{0},R)]\backslash E$. This is certainly true of the multiscale function h above. It is not exactly true of the following examples, but they may be modified slightly to make it so.

Ogata and Tanemura (1981, 1984) borrowed some potential functions ϕ from statistical physics. They give

$$\text{PFI:}\quad h(r)=1+(\alpha r-1)e^{-\beta r^2},\qquad \alpha\geqslant 0,\ \ \beta>0$$

$$\text{PFII:}\quad h(r)=1+(\alpha-1)e^{-\beta r^2}$$

the special case

$$\text{VSC:} \quad h(r) = 1 - e^{-\beta r^2}, \qquad \beta > 0$$

Lennard-Jones:

$$-\ln h(r) = \beta(\sigma/r)^n - \alpha(\sigma/r)^m, \qquad m < n, \quad \beta > 0$$

Other forms suggested by Penttinen (1984) include

$$-\ln h(r) = \theta v[b(\mathbf{0}, R) \cap b(\mathbf{r}, R)]$$

which might be especially appropriate for plant competition.

Ripley and Kelly (1977) called functions h with finite range R (so $h(r) = 1$ for $r > R$) pairwise Markov point processes and characterized them by local properties.

For the rest of this chapter we will assume $d = 2$ where necessary.

4.2 LOCAL CONDITIONING METHODS

Besag (1977a) considered an approximate way to estimate the parameter c of a Strauss process. Note that $c \leqslant 1$ to satisfy the integrability condition (Kelly and Ripley, 1976). Suppose we divide E into a grid of small squares of area Δ. Within each cell we will have a Poisson number n_i of points, distributed uniformly within the cell. The random variables (n_i) are an auto-Poisson process (Besag, 1974). That is,

$$n_i \sim \text{Poisson}(\mu_i)$$
$$\mu_i = \Delta b c^{t_i}$$

where t_i is the total count for cells with at least part within distance R of the centre of cell i. This defines a unique stochastic process on $\mathbb{Z}^2$, and Besag, Milne and Zachary (1982) show rigorously that as $\Delta \to 0$ the process converges weakly to a Strauss process. We have

$$\frac{P(\mathbf{n})}{P(\mathbf{0})} = \prod_{\text{cells}} \frac{(b\Delta)^{n_i}}{n_i!} c^{\frac{1}{2}\Sigma t_i}$$

and hence the cell process has a Radon–Nikodym derivative with respect to the unit Poisson process of

$$\text{const.}\ b^n c^{\frac{1}{2}\Sigma t_i} \qquad \text{on } N(\text{cell}_i) = n_i \forall i$$

It can then be shown that this converges a.s. to const. $b^n c^{Y(R)}$.

We can estimate c for the auto-Poisson process by pseudo-likelihood (PL) (Besag, 1975). Pseudo-likelihood estimation maximizes

$$\text{PL} = \sum_{\text{cells}} P(n_i \mid \text{all } n_j, j \neq i)$$

$$= \prod_i e^{-\mu_i} (\mu_i)^{n_i} / n_i!$$

Thus

$$\ln \mathrm{PL} = -b\Delta \sum_i c^{t_i} + \sum n_i \ln(b\Delta c^{t_i})$$

$$\sim -b \int_E c^{t(\xi)}\, \mathrm{d}\xi + \sum_{\text{pts}} t(\xi) \ln c + n \ln b + n \ln \Delta$$

where $t(\xi)$ denotes the number of points within distance R of $\xi \in E$. This suggests

$$2Y(R) \ln c - b \int_E c^{t(\xi)}\, \mathrm{d}\xi + n \ln b \tag{6}$$

as the appropriate limit of the log pseudo-likelihood as the cells are refined and $\Delta \to 0$. Thus we can estimate b and c by solving $\partial \ln \mathrm{PL}/\partial b = \partial \ln \mathrm{PL}/\partial c = 0$, which gives

$$\left.\begin{aligned} b \int_E c^{t(\xi)}\, \mathrm{d}\xi &= n \\ b \int_E t(\xi) c^{t(\xi)}\, \mathrm{d}\xi &= \sum_{\text{pts}} t(\xi) = 2Y(R) \end{aligned}\right\} \tag{7}$$

Clearly maximizing the exact PL for a fine grid as in Besag (1977a) will give approximately the same estimators.

This argument can be applied quite generally. We find

$$\sum_{\text{pts}} \ln \lambda^*(\xi; \mathbf{x}) - \int_E \lambda^*(\xi; \mathbf{x})\mathrm{d}\xi$$

as the appropriate limiting form of the log pseudo-likelihood.

We must note that here R-close refers to points in $\mathbb{R}^d$, not just within E. One way to avoid edge effects is just to integrate over $E^* \subset E \ominus b(\mathbf{0}, R)$, and so sum over points in E^*. We could also edge-correct $Y(R)$ by the methods of chapter 3.

We can relate (7) to an interesting identity for Gibbsian point processes. Let

$$c_n = \int_{E^n} f(\mathbf{x}) \prod \mathrm{d}\mathbf{x}_i$$

$$p_n = \mathrm{e}^{-a} c_n / n!$$

Then $\mathrm{P}(N = n) = p_n$ and conditional on $N = n$ the points have p.d.f. f/c_n. Now suppose there is a point at ξ, and consider the n remaining points $\mathbf{x}$. Because ξ could be any of the $(n+1)$ points in $\mathbf{x} \cup \{\xi\}$ we find that the conditional Radon–Nikodym derivative is

$$f_\xi(\mathbf{x}) = Cf(\mathbf{x} \cup \{\xi\}) = C\lambda^*(\xi; \mathbf{x}) f(\mathbf{x})$$

where C is chosen to ensure that the derivative integrates to one. This distribution is the so-called reduced Palm distribution because it conditions on the point at ξ but gives the distribution of the remaining points. In the case of a Poisson process of rate b we find $\lambda^* = b$ and hence $C = 1/b$. In general

$$E_\xi(Y) = \mathrm{E}(YC\lambda^*(\xi;\,\mathbf{x}))$$

for any function Y of the process on $E\backslash\{\xi\}$. We can identify C by taking $Y \equiv 1$. Then

$$1/C = \mathrm{E}[\lambda^*(\xi;\,\mathbf{x})] = \lambda(\xi)$$

the intensity at ξ. Finally

$$\mathrm{E}_\xi(Y) = \mathrm{E}[Y\lambda^*(\xi;\,\mathbf{x})]/\lambda(\xi) \tag{8}$$

Now consider a Strauss process and $A = b(\mathbf{0}, R)\backslash\{\mathbf{0}\}$. Then

$$\mathrm{P}_0(N(A) = m) = \frac{bc^m}{\lambda(\mathbf{0})}\,\mathrm{P}(N(A) = m) \tag{9}$$

Let $T = N(A)$ be the number of R-close points to the origin. From (9) we deduce that

$$\mathrm{E}_0 T = \frac{b}{\lambda(\mathbf{0})}\,\mathrm{E}Tc^T, \qquad 1 = \frac{b}{\lambda(\mathbf{0})}\,\mathrm{E}c^T$$

An obvious estimator of $\mathrm{E}_0 Y$ is to measure Y relative to each observed point and to average. Thus

$$\hat{\mathrm{E}}_0 1 = 1$$

$$\hat{\mathrm{E}}_0 T = \frac{1}{n}\sum_{\text{pts}} t(\xi)$$

The equally obvious estimator of $\mathrm{E}[Y\lambda^*(\xi;\,\mathbf{x})]$ is again to average over all points ξ using stationarity. Thus

$$\hat{\mathrm{E}}(c^T) = \frac{1}{a}\int_E c^{t(\xi)}\,\mathrm{d}\xi$$

$$\hat{\mathrm{E}}(Tc^T) = \frac{1}{a}\int_E t(\xi)c^{t(\xi)}\,\mathrm{d}\xi$$

Equating the estimators gives

$$a\lambda(\mathbf{0}) = b\int_E c^{t(\xi)}\,\mathrm{d}\xi$$

$$\frac{a\lambda(\mathbf{0})}{n}\sum_{\text{pts}} t(\xi) = b\int_E t(\xi)c^{t(\xi)}\,\mathrm{d}\xi$$

which recovers (7) if we estimate $\lambda(\mathbf{0})$ by n/a.

Identity (8) was originally described in a much more abstract way by Takacs (1986) and Fiksel (1984). (Despite the dates, Fiksel used a preprint of Takacs' paper dated 1983.) Their definition of a Gibbs process is very abstract, and it is not trivial to show that it is equivalent to that given here. They propose estimating $\lambda^*(\xi; \mathbf{x})$ by fitting (8) for a number of functions Y. We have seen how the left-hand side may be estimated by averaging the view of the process from each point, whereas the right-hand side depends on the parameters and is estimated by stationarity as ξ runs over E. Exactly as for the Strauss process there is a need for edge correction by the methods of chapter 3.

Next consider the multiscale process. The appropriate log pseudo-likelihood is

$$\sum_{\text{pts}} \ln \lambda^*(\xi; \mathbf{x}) - \int_E \lambda^*(\xi; \mathbf{x}) \mathrm{d}\xi$$
$$= \sum_{\text{pts}} \ln b \prod_1^m c_i^{t_i(\xi)} - \int_E b \prod_1^m c_i^{t_i(\xi)} \mathrm{d}\xi$$

where $t_i(\xi)$ is the number of points in $b(\xi, R_i) \backslash b(\xi, R_{i-1})$. Again, finding the stationary point gives

$$b \int_E \sum_1^m c_i^{t_i(\xi)} \mathrm{d}\xi = n$$

$$b \int \prod_1^m c_i^{t_i(\xi)} t_j(\xi) \mathrm{d}\xi = \sum_{\text{pts}} t_j(\xi) \qquad j = 1, \ldots, m$$

generalizing (7) and corresponding to the identities

$$\left.\begin{aligned} \lambda(0) &= \mathrm{E}\left[b \prod_1^m c_i^{T_i} \right] \\ \lambda(0)\mathrm{E}_0 T_j &= \mathrm{E}\left[b \sum_{i=1}^m c_i^{T_i} T_j \right] \qquad j = 1, \ldots, m \end{aligned}\right\} \tag{10}$$

where $T_i = N[b(0, R_i) \backslash b(0, R_{i-1})]$. Takacs (1986) proposed to use (8) in a different way, using the m^2 equations

$$\lambda(0)\mathrm{E}_0 T_i T_j = \mathrm{E}\left[T_i T_j b \prod_{r=1}^m c_r^{T_r} \right] \qquad i, j = 1, \ldots, m$$

and fitting this overdetermined system by least squares to find the $(m+1)$ parameters $(b, c_1, \ldots, c_m)$. (He in fact assumed $c_1 = 0$ and $\lambda(0)$ is known, but the method is easily extended.)

These methods are discussed further in §4.5.

4.3 APPROXIMATE MAXIMUM LIKELIHOOD

The general likelihood of a pairwise interaction process comes from (3);

$$\mathrm{L}(\theta; \mathbf{x}) = a_0(\theta) b(\theta)^{\#\mathbf{x}} \prod_{i<j} h_\theta(d(\mathbf{x}_i, \mathbf{x}_j)) \tag{11}$$

We can also consider conditioning on $N=\#\mathbf{x}=n$ to get

$$\mathrm{L}_n(\theta;\, \mathbf{x})=\frac{1}{Z_\theta}\prod_{i<j} h_\theta(d(\mathbf{x}_i,\, \mathbf{x}_j)) \tag{12}$$

following (5). Note that it is not safe to assume that N is ancillary for θ. For example, in a hard-core model the maximum number of points possible depends strongly on R, and so $p_n=\mathrm{P}(N=n)$ will depend quite strongly on θ. Nevertheless, for computational reasons we might work conditionally on $N=n$.

The problem with both (11) and (12) is that the normalizing constant $a_0(\theta)$ or $1/Z_\theta$ is unknown except via a high-dimensional integral. The approaches described below and in §4.4 either approximate or estimate the normalizing constant.

Poisson approximation

For the Strauss process (12) becomes

$$\mathrm{L}_n(c;\, \mathbf{x})=\frac{1}{Z_\theta}\, c^{t(\mathbf{x})}$$

where $t(\mathbf{x})$ is the number of R-close pairs, previously denoted by $Y(R)$. Now we saw in chapter 3 that for sparse data ($\lambda\pi R^2$ small) from a binomial process $Y(R)$ was approximately Poisson distributed with mean (in $\mathbb{R}^2$) $n(n-1)\pi R^2/2a$. Since

$$Z_\theta=\int_{E^n} c^{t(\mathbf{x})}\cdot\Pi\, \mathrm{d}\mathbf{x}_i=a^n \mathrm{E}c^{Y(R)}$$

this gives the approximation

$$Z_\theta\approx a^n \exp\left\{\frac{n(n-1)}{2a}\,\pi R^2(c-1)\right\}$$

using the probability generating function of a Poisson distribution. Thus

$$\ln\, \mathrm{L}_n(c;\, \mathbf{x})\approx \text{const.}+y(R)\ln c-\frac{n(n-1)}{2a}\,\pi R^2(c-1)$$

and hence

$$\hat{c}\approx y(R)\cdot\frac{2a}{n(n-1)\pi R^2} \tag{13}$$

This idea is due to Penttinen (1984). It can be extended to the unconditional form, for

$$\mathrm{L}(b,\, c;\, \mathbf{x})=a_0(b,\, c)b^{N(E)}c^{Y(R)}$$

so

$$a_0(b,\, c)^{-1}=\mathrm{E}(b^{N(E)}c^{Y(R)})$$

for a Poisson process of unit rate. From the above

$$\mathrm{E}(b^N c^{Y(R)}|N=n) \approx b^n \exp\left\{\frac{n(n-1)}{2a}\pi R^2(c-1)\right\}$$

so

$$\mathrm{E}(b^N c^{Y(R)}) \approx \sum_{n=1}^{\infty} \frac{\mathrm{e}^{-a}a^n}{n!} b^n \exp\left\{\frac{n(n-1)}{2a}\pi R^2(c-1)\right\}$$
$$=\mathrm{e}^{(b-1)a}\mathrm{E}\left[\exp\left\{\frac{N(N-1)}{2a}\pi R^2(c-1)\right\}\right]$$

where $N \sim$ Poisson (ba), so

$$a_0(b, c)^{-1} \approx \mathrm{e}^{ab-a} \exp\left\{\frac{b^2a^2}{2a}\pi R^2(c-1)\right\}$$

and

$$\ln \mathrm{L}(b, c; \mathbf{x}) \approx n \ln b + y(R) \ln c - ab - \tfrac{1}{2}b^2 a\pi R^2(c-1)$$

so

$$\hat{b} \approx n/a, \qquad \hat{c} \approx y(R) \cdot \frac{2a}{n^2\pi R^2}$$

using the sparseness condition when estimating b. Thus in the sparse case we lose little by conditioning.

There are a couple of problems with this approximation. We may find $\hat{c} > 1$, especially if the true c is near one, whereas the Strauss process is only defined for $c \leq 1$. Secondly, edge effects are ignored, which will cause $Y(R)$ to be underrecorded and thus c to be underestimated. As ever, an edge-corrected version can be used.

We can obtain some idea of the variability of $\hat{c}$ at least in the case $c=1$. Then

$$\operatorname{var} \hat{c} = \frac{2a}{n(n-1)\pi R^2} \approx \frac{1}{y(R)}$$

This is also the inverse of the observed information, using the approximations, which suggests its validity for $c<1$ as well. Note that we cannot rely on standard maximum likelihood theory for this statement, since the observations are dependent even for $c=1$. However, the sparse limit case allows us to divide E up into essentially independent subregions, giving some confidence that the classical theory will apply at least approximately.

Thus far we have only considered estimating c. In the hard-core case $(c=0)$ we will wish to estimate R. Ripley and Silverman (1978) noted that the Poisson approximation gave

$$\frac{n(n-1)}{2a}\pi(d_{\min}^2 - R^2) \sim \exp(1)$$

although the maximum likelihood estimator of R is $d_{\min}$. We can use this expression to give confidence intervals for R. In the general case $(0<c<1)$ we can estimate R by maximum likelihood simultaneously with c, although as $y(R)$ is not a continuous function of R we will meet some difficulties and may expect to obtain a biased estimator.

The Poisson approximation can also be used for general pairwise interaction processes (Penttinen, 1984). Suppose that $h(r)=1$ for $r \geqslant R$. Let $d_{(1)}, d_{(2)}, \ldots$ be the ordered interpoint distances. Then we are only concerned with $d_{(1)}, \ldots, d_{(M)}$ where $M=Y(R)$ and

$$Z_\theta = \int_{E^n} \prod_1^{Y(R)} h(d_{(i)}) \prod \mathrm{d}\mathbf{x}_i$$

Under the Poisson approximation

$$Y(R) \sim \text{Poisson}\left(\frac{n(n-1)}{2a}\pi R^2\right)$$

$$d_{(1)}^2, \ldots, d_{(Y(R))}^2 \sim \text{order statistics from } U(0, R^2), \text{ conditional on } Y(R).$$

Thus

$$\begin{aligned} Z_\theta &= \sum_j \mathrm{P}(Y(R)=j) a^n \mathrm{E}\left[\prod_1^j h(\sqrt{U_{i,j}})\right] \\ &\approx a^n \sum \mathrm{P}(Y(R)=j)\left[\frac{2}{R^2}\int_0^R wh(w)\mathrm{d}w\right]^j \\ &= a^n \mathrm{E}\left[\frac{2}{R^2}\int_0^R wh(w)\mathrm{d}w\right]^{Y(R)} \\ &= a^n \exp\left\{\frac{n(n-1)}{2a}\pi R^2\left[\frac{2}{R^2}\int_0^R wh(w)\mathrm{d}w - 1\right]\right\} \\ &= a^n \exp\left\{\frac{n(n-1)}{2a}\int_0^\infty [h(w)-1]2\pi w\,\mathrm{d}w\right\} \end{aligned} \tag{14}$$

As an example, consider the multiscale process.

$$\ln \mathrm{L}_n(\mathbf{c};\, \mathbf{x}) \approx \text{const.} + \sum_1^m [y(R_i) - y(R_{i-1})] \ln c_i - \frac{n(n-1)}{2a}\pi\left[\sum_1^m (R_i^2 - R_{i-1}^2)(c_i - 1)\right]$$

so

$$\hat{c}_i \approx \frac{y(R_i) - y(R_{i-1})}{\pi(R_i^2 - R_{i-1}^2)} \cdot \frac{2a}{n(n-1)}$$

generalizing (13). Again, this result can be generalized to the Poisson case exactly as for the Strauss process. We find that if

$$f(\mathbf{x}) = a_0(\theta) b(\theta)^{N(E)} \prod_{i<j} h_\theta(d(\mathbf{x}_i, \mathbf{x}_j))$$

then

$$a_0(\theta)^{-1} \approx e^{ab-a} \exp\left\{\tfrac{1}{2} b^2 a \int_0^\infty [h(r)-1] 2\pi r \, dr\right\} \tag{15}$$

Approximations from statistical physics

The problem of approximating the partition function Z_θ has a long history in statistical physics. The study of gases concerns very sparse situations so rather crude approximations sufficed. More recent work on the liquid state has produced more refined approximations. These ideas have been borrowed by Ogata and Tanemura (1981, 1984) for use in statistical problems. All their work is conditional on $N = n$.

All the approximations start from so-called *cluster expansions*, which are essentially the calculations used to show that higher moments converge to those of the Poisson in the proofs referenced in §3.4. For ease of notation let $d_{ij} = d(\mathbf{x}_i, \mathbf{x}_j)$ and $f = h - 1$. Then

$$Z_\theta = \int_{E^n} \prod_{i<j} h_\theta(d_{ij}) \Pi \, d\mathbf{x}_i = \int_{E^n} \prod_{i<j} \{1 + f(d_{ij})\} \Pi \, d\mathbf{x}_i$$

Since $h_\theta(r) \to 1$ as $r \to \infty$, usually rapidly, we may expect $f(d_{ij})$ to be small (even zero) for most interpoint distances. If we expand the product we find

$$Z_\theta = \int_{E^n} \{1 + \Sigma f(d_{ij}) + \Sigma\Sigma f(d_{ij}) f(d_{kl}) + \cdots\} \Pi \, d\mathbf{x}_i$$

The successive terms are the clusters of the *Mayer cluster expansion.* To evaluate this expression, view it as a sum over graphs on n sites, the edge from i to j being included if and only if $f(d_{ij})$ occurs in the product. Thus

$$Z_\theta = \sum_{\text{graphs}} \int_{E^n} \prod_{\text{edges}} f(d_{ij}) \Pi \, d\mathbf{x}_i$$

Our expansion will be for large a, n fixed but large (a sparse limit, in which we ignore edge effects). The graphs can be divided into connected components, so if

$$W_{\text{graph}} = \int_{E^k} \prod_{\text{edges}} f(d_{ij}) \Pi \, d\mathbf{x}_i$$

for a connected graph on k points,

$$Z_\theta = \sum_{\text{graphs}} \prod_{\text{components}} W_{\text{component}} a^{n - \#\text{component}}$$

Now all W's are approximately proportional to a, as once one point is fixed, the integrand is non-negligible only for all the others near that point. Thus if $W' = W/a$,

$$Z_\theta = a^n \sum_{\text{graph}} \prod_{\text{components of size} \geq 2} W'_{\text{component}} / (a^{\#\text{component}-1})$$

Collecting together terms of the same power of a gives the expansion. If we ignore powers of a beyond a^{n-4} we find

$$\begin{aligned} Z_\theta = a^n \Bigg[1 &+ \frac{n(n-1)}{2a} W'_{\bullet\!-\!\bullet} \\ &+ \frac{1}{a^2} \left\{ \frac{n(n-1)(n-2)(n-3)}{2 \cdot 2} W'^2_{\bullet\!-\!\bullet} + \frac{n(n-1)(n-2)}{3 \cdot 2} (3W'_{\wedge} + W'_{\triangle}) \right\} \\ &+ \frac{1}{a^3} \left\{ \frac{n(n-1)\ldots(n-5)}{2^3} W'^3_{\bullet\!-\!\bullet} + \frac{n(n-1)\ldots(n-4)}{2 \cdot 3 \cdot 2} W'_{\bullet\!-\!\bullet}(3W'_{\angle} + W'_{\triangle}) \right. \\ &\left. + \frac{n(n-1)(n-2)(n-3)}{4!} (3W'_{\square} + 6W'_{\boxslash} + W'_{\boxtimes} + 12W'_{\sqcup} + 12W'_{Z}) \right\} \Bigg] \end{aligned}$$

On taking logarithms and using $W'_{\angle} = W'^2_{\bullet\!-\!\bullet}$ and so on, we find, for small $n\lambda$,

$$\ln Z_\theta = n \ln a + n \left[\frac{\lambda A_2}{2!} + \frac{\lambda^2 A_3}{3!} + \frac{\lambda^3 A_4}{4!} + O(a^{-5}) \right] \tag{16}$$

where

$$A_k = \int_{\mathbb{R}^{kd}} \sum \prod_{\text{edges}} f(d_{ij}) \prod_2^k \mathrm{d}\mathbf{x}_i$$

the sum being over closed graphs connecting all k sites. If we truncate this expansion at $k=2$, we find

$$\ln Z_\theta \approx n \ln a + \frac{n\lambda A_2}{2} = n \ln a + \frac{n^2}{2a} \int_{\mathbb{R}^d} [h(\|\mathbf{x}\|) - 1] \mathrm{d}\mathbf{x}$$

which is (14) up to the difference between n and $n-1$. A more exact derivation for n fixed and a large gives (14) exactly. The next refinement is

$$\begin{aligned} \ln Z_\theta \approx n \ln a &+ \frac{n^2}{2a} \int_{\mathbb{R}^d} f(d_{12}) \mathrm{d}\mathbf{x}_2 \\ &+ \frac{n^3}{6a^2} \int_{\mathbb{R}^{2d}} f(d_{12}) f(d_{13}) f(d_{31}) \mathrm{d}\mathbf{x}_2 \mathrm{d}\mathbf{x}_3 \end{aligned}$$

Note that since mostly $f \leq 0$, we expect each integral to be negative.

In statistical physics interest lies in derivatives of $\ln Z_\theta$ to give quantities such as the pressure P. It is conventional to use V rather than a for the 'volume' of E. Then

$$P = \lambda k T \left(\frac{\partial \ln Z_\theta}{\partial V} \right) = \frac{nkT}{V} [1 + B_2 \lambda + B_3 \lambda^2 + \cdots] \tag{17}$$

where k is Boltzmann's constant, T is absolute temperature and the coefficients B_k are known as *virial* coefficients. Equation (17) is a correction to Boyle's Law for an imperfect gas. We have

$$B_k = (1-k)A_k/k!$$

which are known for some simple models and small k (up to 7, say). This enables (16) to be used as a better approximation to Z_θ.

Suppose θ contains a scale parameter R. Then

$$A_k(R) = R^{(k-1)d} A_k(1)$$

Let $\tau = nR^d/a$. Then for small τ, with either small R or large a, we find

$$\ln Z_\theta \approx n \ln a + n\left[\frac{\tau}{2!} A_2(1) + \frac{\tau^2 A_3(1)}{3!} + \cdots\right] \tag{16a}$$

Then if

$$\psi(\tau) = -\frac{\tau}{N}\left(\frac{\partial \ln Z_\theta}{\theta\tau}\right)$$

we have

$$\psi(\tau) = B_2\tau + B_3\tau^2 + \cdots \tag{17a}$$

The cases where the virial coefficients are known *are* scale families. For the hard-core model, with $b = \pi/2$,

$$\psi(\tau) = b\tau + 0.782(b\tau)^2 + 0.532\,23(b\tau)^3 + \cdots$$

Note that we do not need this result to find the maximum likelihood estimator of R in a hard-core model, since we saw $\hat{R} = d_{\min}$. However, it is useful for the Strauss process for fixed R. Then if $A_k(c)$ refers to the Strauss process with parameter c,

$$A_2(c) = (c-1)A_2(0)$$
$$A_3(c) = (c-1)^3 A_3(0)$$

but we do not have such a simple expression for $k \geqslant 4$, when different types of graph occur. For example, with $k=4$ we have

with factors

$$(c-1)^4 \qquad (c-1)^5 \qquad (c-1)^6$$

However, we can deduce that

$$\ln Z_c \approx n \ln a + \frac{n^2}{2a}(c-1)\pi R^2 + \frac{n^3 R^4 (c-1)^3}{3!a^2}\, 5.8 \tag{18}$$

The coefficient 5.8 comes from

$$\int f(d_{12})f(d_{23})f(d_{31})d\mathbf{x}_2\, d\mathbf{x}_3$$

$$=(c-1)^3R^4 v\{\mathbf{x}\in b(\mathbf{0},1),\ \mathbf{y}\in b(\mathbf{0},1),\ d(\mathbf{x},\mathbf{y})\leqslant 1\}$$

$$=(c-1)^3R^4\int_0^1 v_B(h)2\pi h\, dh \qquad \text{with } B=b(\mathbf{0},1)$$

since if $\mathbf{x}_2$ is at distance h from the origin, $\mathbf{x}_3$ must lie in $b(\mathbf{0},1)\cap b(\mathbf{x}_2,1)$ which has area $v_B(h)$.

We can extend this to the multiscale process with

$$A_3(\theta)=\int_{\mathbf{R}^4} f(\|\mathbf{x}_2\|)f(\|\mathbf{x}_3\|)f(\|\mathbf{x}_2-\mathbf{x}_3\|)d\mathbf{x}_2\, d\mathbf{x}_3$$

which is the sum of $(c_1-1)^{r_1}(c_2-1)^{r_2}(c_3-1)^{r_3}$ over m^3 subareas, which we can reduce slightly by symmetry. Thus $A_3(\theta)$ is a known but very complex function of θ.

This derivation has been given in some detail as all the accounts the author consulted contain errors. In particular, that of Ogata and Tanemura (1984) is inconsistent and needlessly complicated.

4.4 MONTE-CARLO APPROXIMATIONS

Suppose we wish to estimate θ in h_θ by maximum likelihood. We have

$$\ln \mathrm{L}_n(\theta;\mathbf{x})=\sum_{i<j}\ln h_\theta(d_{ij})-\ln Z_\theta$$

$$\ln \mathrm{L}(\theta;\mathbf{x})=\sum_{i<j}\ln h_\theta(d_{ij})+\ln a_0(\theta)+n\ln b(\theta)$$

To maximize these we need to evaluate the normalizing constants $-\ln Z_\theta$ and $\ln a_0(\theta)$. We end up with evaluating $\mathrm{E}f_0(\mathbf{X})$ for $\mathbf{X}$ from a binomial or Poisson process respectively, f_0 being the known part of f.

The naive Monte-Carlo method would be to sample m patterns $\mathbf{X}_i$ from the base process, and estimate $I=\mathrm{E}f_0(\mathbf{X})$ by

$$\hat{I}=\frac{1}{m}\sum_1^m f_0(\mathbf{X}_i)$$

If f_0 is highly variable (as it usually will be) this can be a very poor estimator of I. Consider the Strauss process on $N(E)=n$. Then

$$\hat{I}=\frac{1}{m}\sum c^{y_i(R)}$$

and if c is small most of the contribution to I will come from the (rare)

Table 4.1

ψ	c	Z_c/a^n	$mCV(\hat{Z}_c)$
5	0.1	1.1×10^{-2}	7.5
	0.5	1.8×10^{-2}	4.8
	0.8	8.2×10^{-2}	1.6
20	0.1	1.5×10^{-8}	3300
	0.5	4.5×10^{-5}	72
	0.8	1.8×10^{-2}	2.1
50	0.1	2.8×10^{-20}	6×10^{8}
	0.5	1.4×10^{-11}	520
	0.8	4.5×10^{-5}	2.5

realizations with small $Y(R)$. We can quantify this via the Poisson approximation. Then

$$\operatorname{var} \hat{Z}_c=\left(\frac{a^{2n}}{m}\right) \operatorname{var} [c^{Y(R)}]$$

$$\approx\frac{a^{2n}}{m} [\exp \psi(c^2-1)-\exp 2\psi(c-1)]$$

where $\psi=n(n-1)\pi R^2/2a$. Table 4.1 shows that if c is small and/or ψ is large then the naive method will be hopelessly inefficient.

An alternative is to generate samples $\mathbf{Y}_i$ from the Gibbs process. Then

$$\mathrm{E}f_0^{-1}(\mathbf{Y})=\mathrm{E}[f(\mathbf{X})/f_0(\mathbf{X})]=a_0$$

but again f_0^{-1} can be very variable. A better idea is to use importance sampling (Ripley, 1987, chapter 5). Suppose

$$f_0(\mathbf{x})=b^{N(E)}c^{Y(R)}$$

and we simulate samples from a Poisson process of intensity λ, rather than one of unit intensity. Then

$$\hat{I}=\frac{1}{m}\sum_1^m\left(\frac{b}{\lambda a}\right)^{N(E)} c^{Y(R)}\,\mathrm{e}^{(\lambda-1)a}$$

is an unbiased estimator of I. If c is near one we might expect $\lambda\approx b/a$ to be a good choice (it is optimal for $c=1$). However, if c is near zero other choices of λ may be more efficient. To illustrate this, consider $c=0$. Then

$$m\hat{I}=\sum_{\text{samples with } y(R)=0} (b/\lambda a)^n\,\mathrm{e}^{(\lambda-1)a}$$

so

$$m^{r-1}\mathrm{E}\hat{I}^r = \mathrm{e}^{(\lambda-1)ar} \sum_n \mathrm{P}(N(E)=n)\mathrm{P}(Y(R)=0 \mid N(E)=n)\left(\frac{b}{\lambda a}\right)^{nr}$$

$$\approx \mathrm{e}^{(\lambda-1)ar} \sum_n \mathrm{e}^{-\lambda a} \frac{(\lambda a)^n}{n!}\left(\frac{b}{\lambda a}\right)^{nr} \exp\left[-\frac{n(n-1)\pi R^2}{2a}\right]$$

using the Poisson approximation. If $a=1$ and $b=50$ we find

λ	30	40	42	45	50	60
$mCV(\hat{I})$	140.7	23.3	21.9	23.4	36.9	229.6

so the optimal value of λ will be a little less than b (and unit λ is extremely inefficient).

To maximize $\ln \mathrm{L}_n$ or $\ln \mathrm{L}$ we do not need to know $\ln Z_\theta$ accurately, only differences for different θ. This is reminiscent of ideas in the design of experiments, and suggests using blocks. For the simulator this means using the same samples for each $\hat{I}_\theta$. We could also use a regression model

$$\ln Z_\theta = A + B\theta + C\theta^2$$

use the same sample points at a set of values of θ, and so estimate B and C. However, B and C are derivatives of $\ln Z_\theta$ and we may as well estimate them directly. Remember

$$f_\theta(\mathbf{x}) = a_0(\theta) f_0(\mathbf{x};\, \theta)$$

where f_0 is completely known. Then the likelihood equation is

$$0 = u_\theta(\mathbf{x}) = \frac{\partial}{\partial\theta} \ln f_\theta(\mathbf{x})$$

$$= \frac{\partial}{\partial\theta} \ln f_0(\mathbf{x};\, \theta) + \frac{\partial}{\partial\theta} \ln a_0(\theta)$$

and

$$-\frac{\partial}{\partial\theta} \ln a_0(\theta) = \frac{\frac{\partial}{\partial\theta}\int f_0(\mathbf{x};\, \theta)\mathrm{d}\mathbf{x}}{\int f_0(\mathbf{x};\, \theta)\mathrm{d}\mathbf{x}} = a_0(\theta)\int \frac{\partial}{\partial\theta} f_0(\mathbf{x};\, \theta)\mathrm{d}\mathbf{x}$$

$$= \mathrm{E}_\theta\left[\frac{\partial}{\partial\theta} \ln f_0(\mathbf{Y};\, \theta)\right]$$

so

$$u_\theta(\mathbf{x}) = \frac{\partial}{\partial\theta} \ln f_0(\mathbf{x};\, \theta) - \mathrm{E}_\theta\left[\frac{\partial}{\partial\theta} \ln f_0(\mathbf{Y};\, \theta)\right]$$

As usual, some regularity conditions are needed. (In the previous model the last term is $B+2C\theta$.) Penttinen (1984) advocates solving $u_\theta(\mathbf{x})=0$ by Newton–Raphson, estimating the mean and mean derivative of the efficient score u by Monte-Carlo. It is perhaps more natural to use Fisher scoring, estimating $\text{var}(u_\theta(\mathbf{Y}))$ by Monte-Carlo. Again, we might use the ideas sketched above to estimate a functional form for $\mathrm{E}_\theta[\partial \ln f_0/\partial\theta]$.

A slight drawback to this approach is that we cannot sample directly from a Gibbsian process, but must use an iterative technique (Ripley, 1987, chapter 4). This will give us a series of correlated samples $(\mathbf{Y}_i)$, and the simulation will need to be analysed carefully using the output analysis of Ripley (1987, chapter 6). This approach is perhaps most appropriate as a check on an approximate value of θ. Given θ, a series of simulations of $\partial \ln f_0/\partial\theta$ will confirm how close $u_\theta(\mathbf{x})$ is to zero, and also suggest a one-step improvement by Fisher scoring.

A further use of Monte-Carlo methods is to estimate the virial coefficients A_k or B_k (Metropolis *et al.*, 1953).

4.5 DISCUSSION

There are a number of problems with these solutions to parameter estimation. Following the discoveries of chapter 2 there has to be some doubt as to the general validity of maximum likelihood estimation. At least in the case of the multiscale process these can be assuaged by general exponential family theory. If $\theta_i=\ln c_i$, $i=1,\ldots,m$, and $\theta_0=\ln b$, we have

$$\mathrm{L}(\boldsymbol{\theta};\mathbf{x})=a_0(\boldsymbol{\theta})\mathrm{e}^{n\theta_0}\prod_1^m \exp\{\theta_i[y(R_i)-y(R_{i-1})]\}$$

General theory then shows that $\ln \mathrm{L}$ is well behaved with a unique maximum, at least if no $y(R_i)-y(R_{i-1})$ is 0 or $\frac{1}{2}n(n-1)$.

There remain two problems. First, all the approximations ignore edge effects except the Monte-Carlo methods. This is a serious disadvantage to the methods based on virial expansions. These are valid for moderate τ but large n, and the calculations of chapter 3 suggest that terms ignored may be as important as those pursued to higher order. Secondly, the maximum likelihood solutions are to the wrong problem. They are for a process defined on E alone, equivalently for a process with no points outside E. Again, this may well be important in practice. Two possible solutions are to consider $E^*\subset E$ and to condition on the points in $E\backslash E^*$, or to consider $E^*\supset E$ and to average over points in $E^*\backslash E$. The Monte-Carlo approach will cope with either, but needs conditional simulations of the Gibbsian process. This can be done by the usual birth-and-death process (Ripley, 1977).

For simplicity let us compare the various methods for the Strauss process. The crudest approximation (the Poisson or first term of the Mayer cluster expansion) gives

$$Z_\theta \approx a^n \exp\left\{\frac{n(n-1)}{2a}(c-1)\pi R^2\right\}$$

and

$$\hat{c} = y(R) \cdot \frac{2a}{n(n-1)\pi R^2}$$

The exact maximum likelihood estimate (MLE) for the process of n points on E is also a function of $y(R)$,

$$\hat{c}_{\text{mle}} = y(R) \Big/ \frac{\mathrm{d}}{\mathrm{d}c} \ln Z_{\hat{c}}$$
$$= \frac{y(R)\hat{c}}{\mathrm{E}_{\hat{c}} Y(R)} \tag{19}$$

so the MLE equates $Y(R)$ to its expectation. The virial expansion gives a more refined approximation to $\mathrm{E}_{\hat{c}} Y(R)$, namely

$$\mathrm{E}_{\hat{c}} Y(R) \approx c\left[\frac{n^2\pi R^2}{2a} + 2.9\frac{n^3R^4}{a^2}(\hat{c}-1)^2\right]$$
$$= \frac{n^2\pi R^2}{2a} c[1 + 1.83\lambda R^2(\hat{c}-1)^2] \tag{20}$$

Note that this is ignoring edge effects. In chapter 3 we saw that for $c=1$

$$\mathrm{E}\, Y(R) \approx \frac{n(n-1)}{2a}\pi R^2\left[1 - \frac{2uR}{3\pi a}\right]$$

Since the limit is for n fixed, $a \to \infty$, the first edge effect term is $O(a^{-1/2})$ but the virial refinement is of order a^{-1}. This seems a fatal flaw of the virial expansion method!

For the unconditional Strauss process on E we find that the likelihood equations reduce to

$$\left.\begin{aligned} n &= \mathrm{E}_{b,c}\mathrm{N} \\ Y(R) &= \mathrm{E}_{b,c} Y(R) \end{aligned}\right\} \tag{21}$$

From (19) we have an approximation to $\mathrm{E}[Y(R)|N=n]$, and so to use (20) we need approximations to $\mathrm{E}N^r$, $r=1,2$ at least. This is only available under sparseness, when we have from the Poisson approximation (§4.2)

$$\mathrm{E}N \approx ab[1 + 2b\pi R^2(c-1)]$$

$$\mathrm{E}Y(R) \approx \frac{c}{2} ab^2\pi R^2$$

The pseudo-likelihood equations are

$$\left.\begin{aligned} n &= b\int c^{t(\xi)}\,\mathrm{d}\xi \\ Y(R) &= \tfrac{1}{2}b\int t(\xi)c^{t(\xi)}\,\mathrm{d}\xi \end{aligned}\right\} \qquad (7)$$

The alternative derivation via Palm probabilities gave

$$\left.\begin{aligned} n &= \frac{nb}{\lambda(\mathbf{0})}\,\mathrm{E}c^{T} \approx ba\,\mathrm{E}c^{T} \\ Y(R) &= \frac{nb}{2\lambda(\mathbf{0})}\,\mathrm{E}Tc^{T} \approx \frac{ba}{2}\,\mathrm{E}Tc^{T} \end{aligned}\right\} \qquad (22)$$

estimating $\lambda(\mathbf{0})$ by n/a. What is the connection between the two methods? Under sparseness

$$\begin{aligned} \mathrm{E}c^{T} &\approx (1-\lambda\pi R^2)\times 1+\lambda\pi R^2\times c \\ &= 1+\lambda\pi R^2(c-1) \end{aligned}$$

Similarly,

$$\mathrm{E}Tc^{T} \approx \lambda\pi R^2 \times c$$

so at this level of approximation the likelihood and pseudo-likelihood equations agree almost exactly. It seems unknown to what extent this agreement holds at higher interactions.

The local conditioning approach has the advantage of being rather easy to correct for edge effects by the border method, and more efficient corrections may be possible. Since the calculations above suggest that it may be quite efficient (statistically), it would appear to be the method of choice.

4.6 AN EXAMPLE

Ripley (1977) considered the pattern of 69 towns in a 40×40 mile square in Spain, data from Glass and Tobler (1971). The pattern is illustrated in figure 4.1. It was suggested by second-moment means (such as $\hat{K}$ plots) that there was inhibition between the towns up to a distance of $R=3.5$ miles. Ripley (1977) fitted a Strauss model with $\hat{c}=0.5$ by trial-and-error comparison of $\mathrm{E}\hat{K}(\ )$ with the observed $\hat{K}(\ )$, using isotropic edge correction. Figure 4.2 illustrates the process.

Besag (1977) used his approximating auto-Poisson process to suggest $\hat{c}=0.83$ by pseudo-likelihood. This (apparently) ignored edge effects and seems from figure 4.2 to fit the second-moment behaviour less well than $\hat{c}=0.50$.

Figure 4.1 Plot of the locations of 69 towns in a 40 mile square region in the Spanish plain. (Redrawn with permission from Ripley, 1977, figure 13).

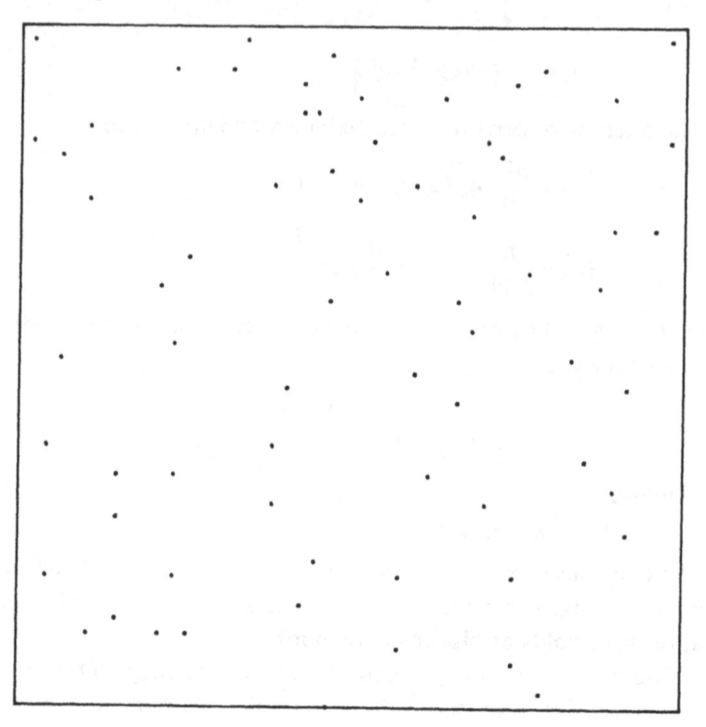

We can compare this with using (7), both with and without edge correction. Without edge correction $Y(R)=30$. Using a border of width R gives an estimate of $2Y(R)/n$ of 0.894 against 0.870 for the uncorrected version. Figure 4.3 plots $\int t(\xi)c^{t(\xi)}\,d\xi/\int c^{t(\xi)}\,d\xi$ against c with and without edge corrections. The estimators obtained were $\hat{c}=0.40$ and $\hat{c}=0.44$ respectively. Isotropic edge correction gives $2Y(R)/n=0.922$ (an increase of 6%) and $\hat{c}=0.41$. We can also use (22), estimating the right-hand sides from simulations of a Strauss process. Since T is the number of points R-close to the origin, we only need to simulate the process on a large enough set to avoid edge effects. Figure 4.4 shows the estimate of $\mathrm{E}Tc^T/\mathrm{E}c^T$ based on 5000 steps of a spatial birth-and-death process. Note that the individual estimates are quite variable.

The sparse data approximation is $\hat{c}=0.53$, with estimated standard deviation of $1/\sqrt{Y(R)}=0.18$. Using the virial expansion at (20) and solving $\mathrm{E}_{\hat{c}}Y(R)=Y(R)$ with no edge correction gives $\hat{c}=0.45$, the second term of (20) increasing $\mathrm{E}_c Y(R)$ by about 9%, a comparable effect to edge correction.

Figure 4.2 Plots of $\sqrt{\hat{K}_2|\pi}$ for the Spanish towns ($\times$) together with the average of 100 simulations of a Strauss process with $R=3.5$ miles. (a) $c=1$. (b) $c=0.8$. (c)$=0.5$. (d) $c=0.3$.

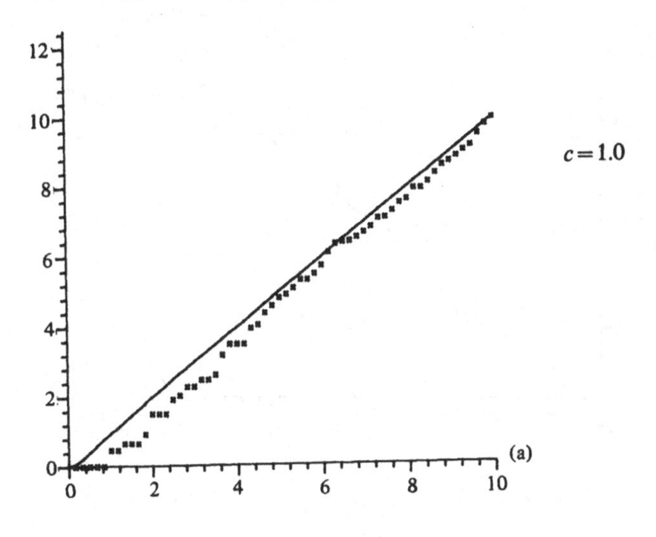

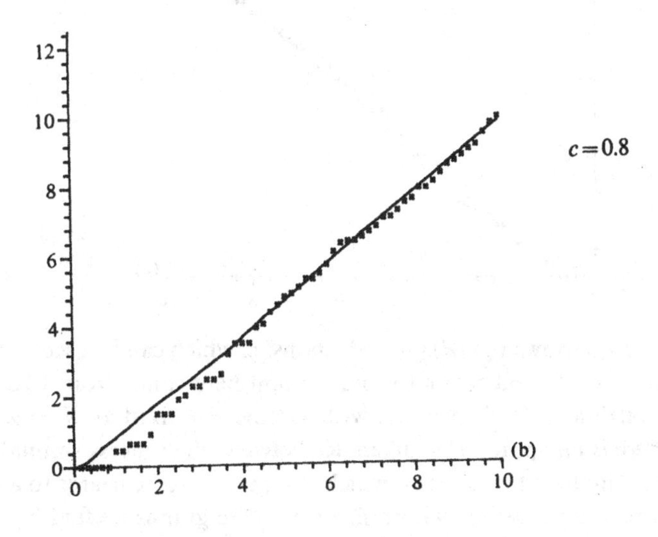

Figure 4.2 Continued.

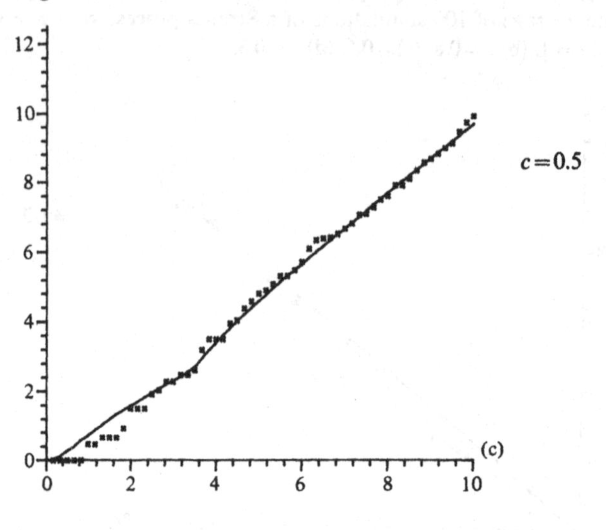

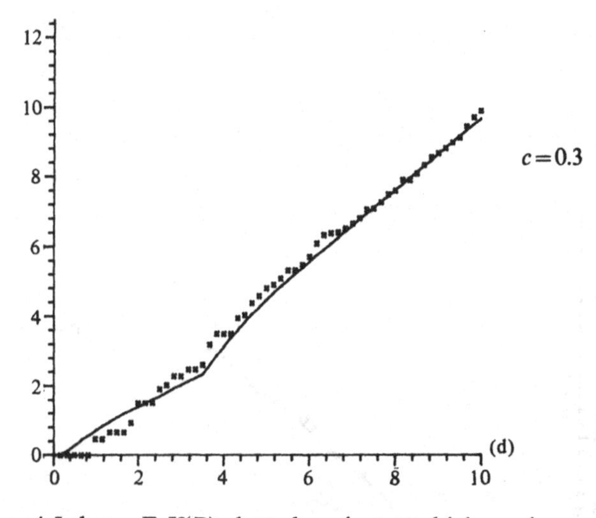

Figure 4.5 shows $\mathrm{E}_c Y(R)$ plotted against c, which can be used to find the maximum likelihood estimator. Each point here comes from 10 000 steps of the birth-and-death process, with sample size fixed at $N=69$. Notice that there is an appreciable difference between the process simulated in E and that simulated on a torus, which is supposed to be nearer to a process throughout $\mathbb{R}^2$ observed within E. From these graphs we find $\hat{c}=0.49$ for the process within E (consistent with the virial expansion), and $\hat{c}=0.44$ for the process on $\mathbb{R}^2$.

Figure 4.3 Plot of $\int t(\xi)c^{t(\xi)}\,d\xi / \int c^{t(\xi)}\,d\xi$ against c for the Spanish towns. The solid line is estimated from the whole data, the dashed line with a border of size R.

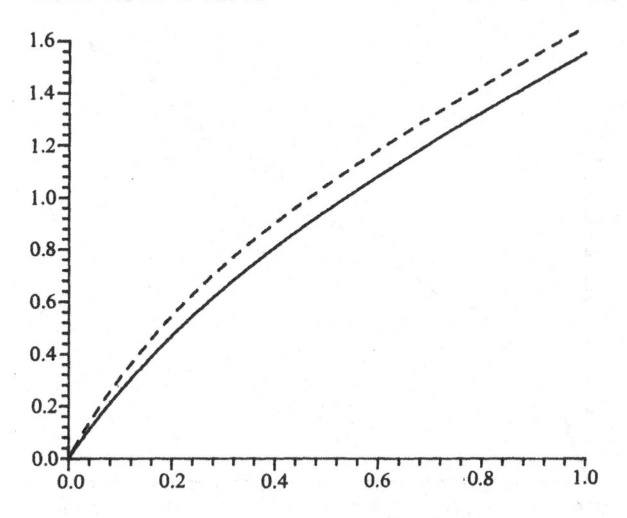

Figure 4.4 $\mathrm{E}Tc^T/\mathrm{E}c^T$ estimated by simulation.

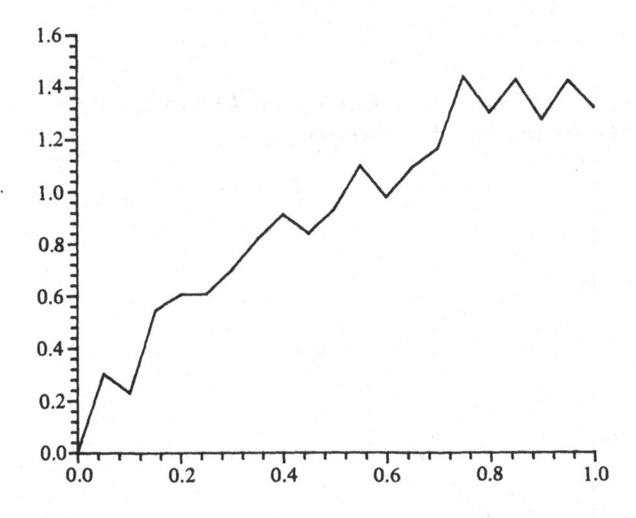

Figure 4.6 shows how we can make more use of the Monte-Carlo results. The linear regression is

$$\mathrm{E}_c Y(R) = 11.8 + 40.8c$$

for the process on $\mathbb{R}^2$ observed in E. Thus $\hat{c} = 0.45$. Further, the standard error in this formula is 0.12 and internal evidence (by use of batches of size

Figure 4.5 Plot of $E_c Y(R)$ against c estimated by simulation. The solid line is without edge correction, the dashed line with toroidal edge correction.

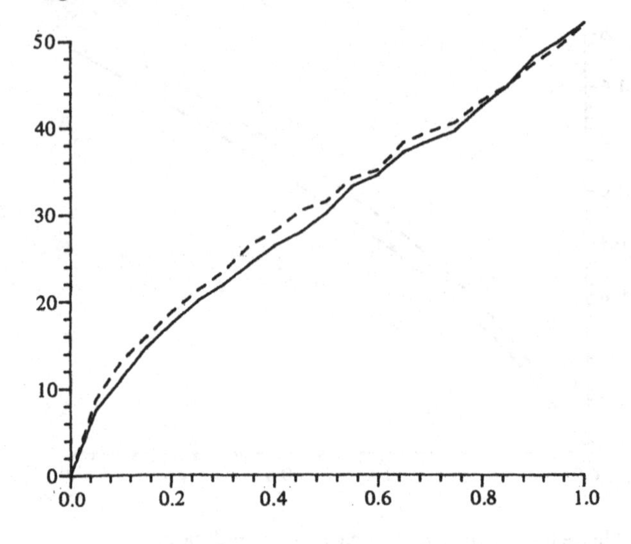

Figure 4.6 An enlarged part of figure 4.5 for toroidal edge correction. The line was fitted by regression.

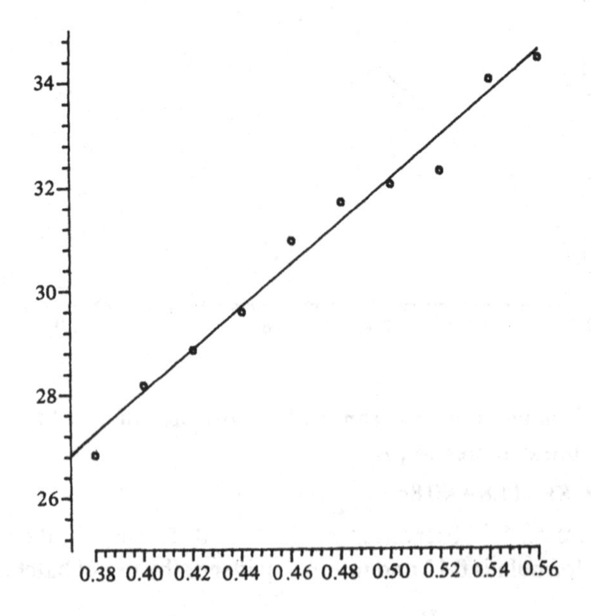

5

Modelling spatial images

Images as data are occurring increasingly frequently in a wide range of scientific disciplines. The scale of the images varies widely, from meteorological satellites which view scenes thousands of kilometres square and optical astronomy looking at sections of space, down to electron microscopy working at scales of $10\,\mu m$ or less. However, they all have in common a digital output of an image. With a few exceptions this is on a square grid, so each output measures the image within a small square known as a *pixel*. The measurement on each pixel can be a greylevel, typically one of 64 or 256 levels of luminance, or a series of greylevels representing luminance in different spectral bands. For example, earth resources satellites use luminance in the visual and infrared bands, typically four to seven numbers in total. One may of course use three bands to represent red, blue and green and so record an arbitrary colour on each pixel.

The resolution (the size of each pixel, hence the number of pixels per scene) is often limited by hardware considerations in the sensors. Optical astronomers now use 512×512 arrays of CCD (charge coupled device) sensors to replace photographic plates. The size of the pixels is limited by physical problems and also by the fact that these detectors count photons, so random events limit the practicable precision. In many other applications the limiting factor is digital communication speed. Digital images can be enormous in data-processing terms. One scene from a LANDSAT satellite contains 30 Mbytes of data yet covers a scene only 185 km square. The need to transmit data back to earth limits the desirable resolution. In real-time applications such as computer vision the need to process the image again limits the resolution. In practice this means that most images available today are used in 64×64 to 512×512 sections. These limits will increase quite rapidly with time.

1000) gives (for c in this range)

$$\text{var } Y(R) \approx 5.9$$

which corresponds to

$$\text{st. dev.}(\hat{c}_{\text{mle}}) \approx 0.14$$

The summary of these studies is that pseudo-likelihood comes out quite well, providing values in close agreement with maximum likelihood at about one-hundredth of the cost. Use of Monte-Carlo for approximate MLEs *is* quite feasible; figure 4.6 took about 3 minutes of CPU time on a VAX 8600.

We might finally want to check the adequacy of the Strauss model by overfitting a multiscale model. With $R_i = 1, 2, 3, 4, 5$ miles we found (by pseudo-likelihood)

range=	0–1	1–2	2–3	3–4	4–5	miles
c_i	0.16	0.56	0.50	0.98	1.24	

which accords well with the fitted Strauss model but has some suspicion of smaller c at very short distances. Note however that this last figure is based on the very small number of pairs less than one mile apart (expected value 2.1 if $c = 0.45$) and so $\hat{c}_1$ is highly variable. Using the Fisher information from the sparse data approximation (surely valid here) gives st.dev.$(\hat{c}_1) \approx 0.7$ under the fitted Strauss model. The variability of the estimators in the multiscale process is always a problem except for very large sample sizes.

Terms such as image processing and image analysis are used to cover quite a range of aims in processing image data. Until recently most of the work has been statistically simple. Major considerations have been the 'cleaning' of images, by removing sensor distortions and smoothing out noise. This has been primarily to ease the job of human interpreters, so this activity is called *image enhancement*. A specific example considered in more detail in §5.3 is to correct astronomical pictures for atmospheric blurring. Measurements on enhanced images are often restricted to simple counting and computing areas; the complex pattern recognition operations have been left to humans.

A further phase of the processing is *pattern recognition*, automating some of the tasks of a human image interpreter. Automatic methods benefit from image cleaning as much as human ones, so the preprocessing phase is still needed. A very simple task might be to consider images such as figure 5.1. This was produced under carefully controlled lighting conditions by digitizing the output from a cheap TV camera. (Even under these conditions there are some problems with specular reflections.) A simple task in computer vision would be to identify the objects present. These were in fact part of a DIY assembly kit for a piece of furniture, so a robot would need this information to begin assembly.

A few more terms need some definition. *Image restoration* is often used

Figure 5.1 A 256 × 256 digital binary image of parts of a DIY assembly kit on a white background. (Reproduced with permission from Ripley and Taylor, 1987.)

to indicate the de-blurring phase of image enhancement. *Segmentation* of images denotes both their division into homogeneous subimages and labelling pixels by their type. For example, in a satellite image we might wish to label various land uses by different colours on a map. One particular type of pattern recognition is *symbol recognition*, both reading continuous text and identifying isolated symbols on a map.

One of the innovations of the last five years has been the success of more statistically-based methods in this field. These methods are based on formal statistical models, hence the title of this chapter. It is difficult to attribute many of the ideas discussed below, which have been developed concurrently by a group of people in loose collaboration. The vision for this approach belongs to Ulf Grenander whose ideas have been seminal (e.g. Grenander, 1983).

5.1 A GENERAL BAYESIAN APPROACH

From now on we will suppose that our image is made up of an $M \times N$ rectangular array of pixels, and that Z_{ij} denotes the measurement at the (i, j) pixel. This measurement may be scalar (a greylevel) or a vector. Corresponding to Z_{ij} there is a true signal S_{ij} representing what would be measured under ideal conditions. This will certainly be contaminated by noise and may also be blurred. Thus if **S** and **Z** represent vectors of signals of measurements,

$$\mathbf{Z} = H(\mathbf{S}, \boldsymbol{\varepsilon}) \tag{1}$$

for a noise vector $\boldsymbol{\varepsilon}$, assumed throughout to be independent of **S**. The problem of *restoration* is to infer **S** from **Z**. In that sense this is an inverse problem and there is a whole class of nonstatistical methods to tackle such problems. Usually we can simplify (1) to

$$Z_{ij} = \Sigma h_{rs} S_{i+r,j+s} + \varepsilon_{ij} \tag{2}$$

in which the 'blurring' is a convolution and the noise is additive. Then in matrix terms

$$\mathbf{Z} = H\mathbf{S} + \boldsymbol{\varepsilon} \tag{2a}$$

For *segmentation* we assume that S_{ij} takes one of a finite number of values $\boldsymbol{\mu}_1, \ldots, \boldsymbol{\mu}_c$ corresponding to the c possible types of pixel. Let l_{ij} denote the type (*label*) of pixel (i, j). The task is then to infer **l** from **Z**.

Thus far we have said very little about the noise present. We may need to allow the (ε_{ij}) to be spatially correlated. Further, the variability of ε_{ij} might depend on l_{ij}. For example, the 'colour' of a wheat field may vary from pixel to pixel both due to local agricultural conditions and to lighting conditions. This random effect is subsumed into ε_{ij}, whose variability we might expect to depend on the land use. (Water is much less variable than wheat, for example.)

Image problems differ from most other spatial problems in that they do not occur in isolation. Almost invariably there will be many images to be processed which were obtained under similar conditions. Thus we will have replication, not of the signal **S**, but of the noise ε and blurring function H. It may well be realistic to regard both H and the distribution of ε as completely known, or known up to a small number of parameters. For example, in studies of LANDSAT data a model of ε has been proposed which seems quite widely applicable. This gives ε_{ij} a multivariate normal distribution with a known covariance matrix Σ, and some simple local correlation between the ε_{ij}'s. In digital astronomy much of the noise comes from the discrete nature of photons, and so Poisson noise may be appropriate, independent from pixel to pixel. The blurring may be known from physical considerations, or it may be known from past experience. (Astronomers know a lot about atmospheric distortion.)

The crucial steps in our general approach are to assume that we have a model for **S** and a model for the distortion of **S** to **Z**. Then we can write

$$P(\mathbf{S}|\mathbf{Z}) \propto P(\mathbf{Z}|\mathbf{S})P(\mathbf{S}) \tag{3}$$

in the usual Bayesian way, and base inference on $P(\mathbf{S}|\mathbf{Z})$. In a segmentation problem **S** will be in a one–one relation with **1**, so we can replace **S** by **1** in (3). There are immediate practical difficulties (these vectors have dimension MN, perhaps 250 000) which we will postpone. It will be impossible to look at the left-hand side of (3) as a function of **S**, so we consider its mode. The so-called MAP technique (Maximum A Posteriori) chooses **S** to maximize $P(\mathbf{S}|\mathbf{Z})$. Equivalently, we choose **S** to minimize

$$\begin{aligned} &-2 \ln P(\mathbf{Z}|\mathbf{S})P(\mathbf{S}) \\ &= -2 \ln P(\mathbf{Z}|\mathbf{S}) - 2 \ln P(\mathbf{S}) \end{aligned}$$

If we use a Gibbsian model for **S**, we will have

$$P(\mathbf{S}) \propto \exp[-\beta U(\mathbf{S})]$$

where $U(\mathbf{s})$ is the 'energy' we attribute to the signal **s**. We will give low energy to signals which agree with our prior conceptions about the image, high energy to those which do not. Thus the energy is a measure of 'roughness' if our prior ideas are that true images are smooth. If we assume (2a) and independent noise we find

$$I = \sum_{\text{pixels}} \Delta_\Sigma(Z_{ij}, HS_{ij})$$

where Δ_Σ denotes Mahalanobis distance with respect to the covariance matrix Σ. Thus I is a measure of the (in)fidelity of the observations **Z** to the signal S. If Σ is known only up to a scale factor, say $\Sigma=\kappa\Sigma_0$ then

$$I=\frac{1}{\kappa}I_0=\frac{1}{\kappa}\sum\Delta_{\Sigma_0}(Z_{ij},HS_{ij})$$

and the MAP estimator minimizes

$$I_0+\lambda U \tag{4}$$

where $\lambda=\kappa\beta$. Thus

MAP minimizes infidelity + λ roughness

This interpretation holds quite generally. The reader will recognize (4) as a Lagrangian form, so MAP also solves

min I subject to $U\leqslant c_1$

min U subject to $I\leqslant c_2$

Further, the form of (4) has been suggested directly as a penalized likelihood (Good and Gaskins, 1971). Special cases have been suggested in other fields. One which has a wide following is *maximum entropy* (see, for images, Skilling and Gull, 1985). This maximizes an entropy E subject to an infidelity constraint. Since high entropy indicates smoothness, we can regard $-E$ as a roughness. Details of the implied choices of U and I are discussed in §5.3.

These alternative approaches may placate those who are not committed to a Bayesian point of view. However, many who doubt the universal validity of the Bayesian paradigm accept it in situations such as this where there is a sequence of related decisions to be made, and quite strong prior information is available from past experience. This is objective rather than subjective Bayesian inference.

In the segmentation problem the MAP solution emerges from a formal decision theory problem. Suppose we set up a loss function which is 1 for all incorrect maps but 0 for the correct map. This leads to selecting the map with the highest *a posteriori* probability as the Bayes rule, and is a widely used paradigm in statistical pattern recognition (Devijver and Kittler, 1982). Note the distinction between the approach so far and that of choosing l_{ij} to maximize $P(l_{ij}|Z)$ which corresponds to the loss function of the number of classification errors made. This is discussed further in §5.5.

Before considering examples in §5.3 and §5.4 we need to tackle the problems of specifying U and computing MAP estimators. There is also the problem of deciding on λ (equivalently, on c_1 or c_2). A number of approaches are possible:

(a) *Parameter estimation.* We know $\lambda=\beta\kappa$. One or other may be known and both can be estimated.

(b) *Cross-validation.* Regard λ as a parameter in a smoothing procedure (Titterington, 1985), and choose it to achieve some measure of a 'good' solution.

(c) *Past experience.* Use a λ which worked well in similar problems in the past.

These are all subjects of current research.

5.2 MARKOV RANDOM FIELD MODELS

We need a probability distribution over possible signals **S** or maps **l** which reflects our prior ideas about the undistorted image. For computational reasons it is highly desirable to have a *local* description of the probability model. Partly this is so that the MAP estimate can be computed locally, and partly so that the complexity of the model does not increase more than linearly with MN, the number of pixels. Markov random fields (Kinderman and Snell, 1980) are ideal candidates, and were exploited by Geman and Geman (1984). For a finite collection of sites (pixels here), a Markov random field (MRF) is a distribution of a collection of random variables X_i such that

$$\mathrm{P}(X_i \mid X_j, j \neq i) \text{ depends only on } \{X_j \mid j \text{ a neighbour of } i\}$$

The neighbour structure can be specified by a graph, with sites being connected by an edge if and only if they are neighbours. The commonest choices for our rectangular array of sites are the 4-neighbour graph (horizontal and vertical adjacencies only) and the 8-neighbour (including diagonals). These have rather different implications for the prior notions of smoothness, as will be exemplified below. Two examples of such models are:

(a) CAR model. This is described in §2.1 and has

$$\mathbf{X} \sim MVN(\boldsymbol{\mu}, \kappa(I - \phi N)^{-1})$$

where N is the neighbour incidence matrix (so $N_{ij} = 1$ if i and j are neighbours, 0 otherwise). For ϕ near its upper bound (the reciprocal of the number of neighbours) this has realizations which appear to come from smoothly varying surfaces. As such it may be a suitable model for **S** in some restoration problems, and is used in §5.3.

(b) Strauss (1977) model. This is a model for unordered categories, such as labels on a map. There $U(\mathbf{l})$ is the number of pairs of neighbouring pixels with dissimilar labels. Thus the most probable maps are all of one colour, and the least probable ones (for 4-neighbours) are patchwork quilts. This *is* Markov, since

$$\mathrm{P}(l_{ij}=k \mid \text{rest of the } l\text{'s}) \propto \exp[\beta \,\#\, (\text{neighbours of colour } k)] \quad (5)$$

The model is exchangeable in the labels, but can be extended to allow different marginal probabilities of each label. In the particular case $c=2$ with the 4-neighbour graph this reduces to the well-known Ising model of statistical physics. This is known to have a critical value of β, $\beta_c = \sinh^{-1} 1 \approx 0.88$ such that, asymptotically as $MN \to \infty$, for $\beta < \beta_c$ there are no infinite patches of one type whereas for $\beta > \beta_c$ there will always be such infinite patches. This means that for $\beta > \beta_c$ the realizations of a large but finite portion of the process will probably be dominated by one colour. We can see this from the definition. Suppose we have a $N \times N$ square grid and colours black and white. The highest probability maps are the two with $U=0$, all black and all white. Next are maps with single isolated pixels. There are 8 such with $U=2$ (pixels at the corners), $4(N-1)$ with $U=3$ and $2(N-2)^2$ with $U=4$. The most probable maps not of isolated clumps of one colour in a sea of the other are those with a straight boundary across the image. These have $U=N$, and there are $4(N-1)$ such maps. Since each of these has probability $\exp(-4\beta N)$ times that of just one colour, they would not appear to be very probable as a class. One would have to add in other less straight boundaries for a more comprehensive picture. As a simple example, consider a 4×4 grid. Even here there are $2^{16} = 65\,536$ possible maps. We have the counts in table 5.1. We will see later that values of β around 2 are plausible. In that case a line across has probability about 1/150 that of a map all of one colour. The most probable

Table 5.1. *Some configuration counts for a 4×4 Ising model*

U	Number	Description
0	2	all one colour
2	8	corner different
3	24	edge different
3	16	□□ at corner
4	4	isolated pixel
4	8	□□ at edge
4	24	two corners different
4	12	straight line across
5	48	line across of length 5
6	112	line across of length 6
7	56	line across of length 7 (max. length)

classes of maps are those of one colour and then those with an isolated small patch of one colour.

Much more needs to be learnt about these simple models. They are very easy to simulate by iterative methods (Ripley, 1987, §4.7), and much can be learnt by watching simulations.

The simple models are not enough to express some ideas of smoothness. For example, in modelling a landscape we will be quite happy with long straight boundaries between classes which will be unduly penalized in the Strauss model. Geman and Geman (1984) introduced the idea of an *edge process*. The graph is modified as shown in figure 5.2. Pixels are not neighbours if the edge element between them is 'on'. At the expense of a

Figure 5.2 A grid of pixels with an edge process (thick lines). Pixels with an edge between them are deemed to be not adjacent. The edge process continues, turns left or right, has a T or a + junction or stops at each corner.

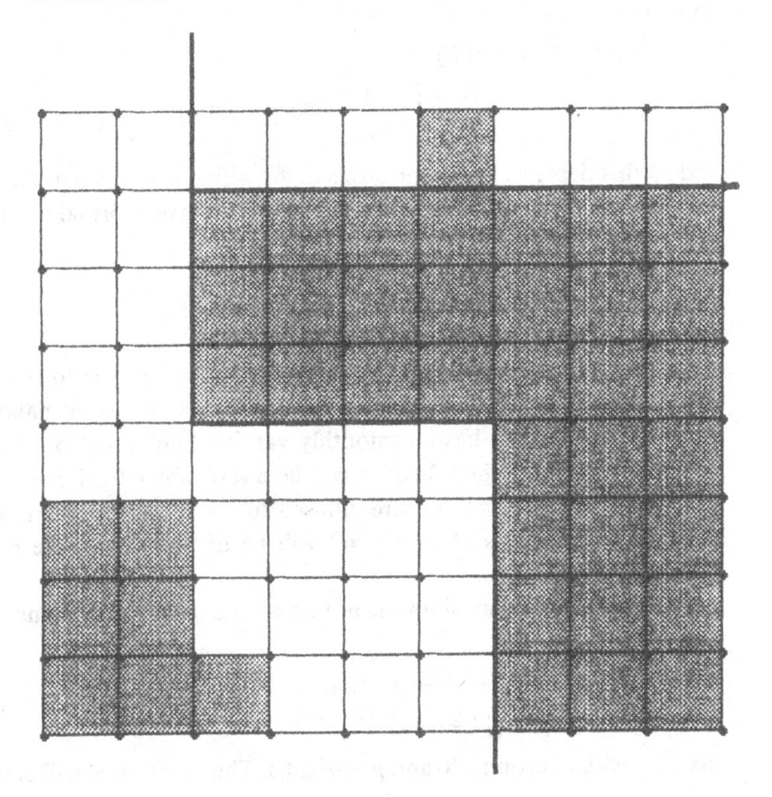

more complicated but still local model one can express more sophisticated prior knowledge. Wohlberg and Pavlidis (1985) use a different model with similar aims, their model using neighbours with a 5×5 or 7×7 neighbourhood. (Their description is rather confusing. The conditional value at a pixel will depend on the values of all pixels in the neighborhood stated here.) Other specifications will doubtless be developed in special applications.

It is important for the optimization algorithms discussed below that $P(\mathbf{S}|\mathbf{Z})$ is also a Markov random field on the same or a similar graph. This will be so when $\boldsymbol{\varepsilon}$ is a Markov random field and H is defined locally. Formal proofs are rather cumbersome, but the assertion and the appropriate graph are usually obvious in an example. Consider (2a) with independent noise $\boldsymbol{\varepsilon}$ and a CAR process for $\mathbf{S}$. Then

$$P(\mathbf{S}|\mathbf{Z}) \propto \exp\left[-\frac{1}{2\kappa_n}\Sigma(Z_{ij}-HS_{ij})^2-\frac{1}{2\kappa_s}\mathbf{S}^{\mathrm{T}}(I-\phi N)\mathbf{S}\right]$$

$$=\exp\left[-\frac{1}{2\kappa_n}\{\mathbf{Z}^{\mathrm{T}}\mathbf{Z}-2\mathbf{Z}^{\mathrm{T}}H\mathbf{S}+\mathbf{S}^{\mathrm{T}}H^{\mathrm{T}}H\mathbf{S}\}-\frac{1}{2\kappa_s}\mathbf{S}^{\mathrm{T}}(I-\phi N)\mathbf{S}\right]$$

and so

$$P(S_{ij}|\mathbf{Z}, \text{rest of } \mathbf{S})$$

$$\propto \exp\left[-\frac{1}{\kappa_n}\{H^{\mathrm{T}}H\mathbf{S}-H^{\mathrm{T}}\mathbf{Z}\}_{ij}-\frac{1}{\kappa_s}\{(I-\phi N)\mathbf{S}\}_{ij}\right]S_{ij}$$

and so this depends on S_{kl} for pixels kl for which kl is a neighbour of ij or $(H^{\mathrm{T}}H)_{ij,kl} \neq 0$. Geman and Geman (1984, p. 728) give a formal proof which generalizes this idea.

5.3 APPLICATIONS IN ASTRONOMY

A typical problem in digital optical astronomy is to remove the effects of atmospheric blurring from an image of a distant galaxy. The galaxy is assumed to have a smoothly varying luminosity, but there will probably be stars in the field of view. These stars are in truth point sources (smaller than one pixel) but are diffused to bright peaks by atmospheric diffusion. The counts at each pixel will be of photons in the range 0–10 000.

The presence of stars allows H to be assessed. It has been found that the function

$$H_{ij,kl}=h(\sqrt{\{(i-k)^2+(j-l)^2\}})$$
$$h(r)=\beta/\pi r_0^2[1+(r/r_0)^2]^\beta$$

fits well, with r_0 around 20 and β around 3. The exact value will depend on the atmospheric conditions that night and the length of the exposure. The values of β and r_0 can be estimated by fitting point sources to isolated stars. Formally, the model for **S** has two independent components, a marked Poisson process of stars (the marks denoting luminances) and a continuous process for the luminosity of the galaxies.

The observation process may be modelled by assuming that the photon count Z_i at detector i is Poisson with mean $(H\mathbf{S})_i$. The Poisson distribution comes from the physics of the emission of photons. We are here making a further assumption of the uniform sensitivity of the detectors to photons. This is known to be false, but the data **Z** are adjusted (multiplicatively) for known variations in the array of detectors as part of the instrument calibration procedure.

The full problem is too complex to treat in an introduction. Here we ignore the possible presence of stars and replace Poisson noise by additive Gaussian white noise of constant variance κ_n. (Closer approximations are considered later.) The photon counts are large, typically 200–10 000, so we lose little by assuming that Z_i is continuous. Since background luminosity typically corresponds to counts of a few hundred photons, the constraints $\mathbf{Z} \geqslant 0$ and $\mathbf{S} \geqslant 0$ which we ignore prove to be automatically satisfied. Thus our simplified model is

$$\mathbf{Z} = H\mathbf{S} + \boldsymbol{\varepsilon} \qquad \boldsymbol{\varepsilon} \sim \text{white noise } (\kappa_n)$$

$$\mathbf{S} \text{ satisfies a CAR } (C, \kappa_s), \text{ mean } \mathbf{0}$$

In assuming a CAR process for S we lose no generality, as *any* Gaussian distribution on a finite set of sites can be expressed as a CAR (Ripley, 1981, p. 90). Nevertheless, we should think of C as local such as ϕN above. Then

$$-2 \ln \mathrm{P}(\mathbf{S}|\mathbf{Z}) = \text{const.} + \frac{1}{\kappa_n} \|\mathbf{Z} - H\mathbf{S}\|^2 + \frac{1}{\kappa_s} \mathbf{S}^{\mathrm{T}}(I - C)\mathbf{S}$$

has a minimum at the stationary point given by

$$(H^{\mathrm{T}}H + \lambda I - \lambda C)\hat{\mathbf{S}} = H^{\mathrm{T}}\mathbf{Z} \qquad (6)$$

where $\lambda = \kappa_n/\kappa_s$ has the same role as in (4). The solution with $\lambda = 0$ is the least squares fit of **S** to **Z** corresponding to vague prior knowledge. The general solution to (6) is often known as a *regularization* of the least squares solution. Such procedures have a long history and a large literature (see, e.g., Frieden, 1979).

Equations (6) form a large but sparse system of linear equations for $\hat{\mathbf{S}}$. Their size (up to $512^2 \times 512^2$) precludes conventional methods of solution, but several iterative methods can be used. In the absence of blurring (6) reduces to

$$\hat{\mathbf{S}} = \alpha\mathbf{Z} + (1 - \alpha)C\hat{\mathbf{S}} \qquad (7)$$

where $\alpha = \kappa_s/(\kappa_n + \kappa_s)$. The $\hat{\mathbf{S}}$ is a convex combination of the observations and a local average of $\mathbf{S}$. This equation provides the iterative scheme

$$\hat{\mathbf{S}}^{(i)} = \alpha \mathbf{Z} + (1-\alpha) C \hat{\mathbf{S}}^{(i-1)}$$

which we will show is usually convergent. With blurring we have

$$H^{\mathrm{T}}(\mathbf{Z} - H\hat{\mathbf{S}}) = \lambda(I - C)\hat{\mathbf{S}}$$

so

$$\alpha H^{\mathrm{T}}(\mathbf{Z} - H\hat{\mathbf{S}}) = \alpha\lambda(I - C)\hat{\mathbf{S}} = (1-\alpha)(I - C)\hat{\mathbf{S}}$$

and

$$\begin{aligned} \hat{\mathbf{S}} &= \hat{\mathbf{S}} + \alpha H^{\mathrm{T}}(\mathbf{Z} - H\hat{\mathbf{S}}) - (1-\alpha)(I-C)\hat{\mathbf{S}} \\ &= (1-\alpha)C\hat{\mathbf{S}} + \alpha[H^{\mathrm{T}}(\mathbf{Z} - H\hat{\mathbf{S}}) + \hat{\mathbf{S}}] \end{aligned} \quad \textbf{(7a)}$$

which reduces to (7) in the absence of blurring. This again gives a convergent iterative scheme. We can rewrite (7a) as

$$\hat{\mathbf{S}} = \alpha H^{\mathrm{T}}\mathbf{Z} + U\hat{\mathbf{S}}$$

where U is the symmetric matrix $(1-\alpha)C + \alpha(I - H^{\mathrm{T}}H)$, with iteration

$$\mathbf{s} \leftarrow \alpha H^{\mathrm{T}}\mathbf{Z} + U\mathbf{s}$$

It suffices to prove that U is a contraction mapping, which we will do by showing that all its eigenvalues have modulus strictly less than one. Consider

$$\begin{aligned} \mathbf{s}^{\mathrm{T}}U\mathbf{s} &= (1-\alpha)\mathbf{s}^{\mathrm{T}}C\mathbf{s} + \alpha\|\mathbf{s}\|^2 - \alpha\|H\mathbf{s}\|^2 \\ &\leqslant (1-\alpha)\lambda_n\|\mathbf{s}\|^2 + \alpha\|\mathbf{s}\|^2 \\ &\leqslant [(1-\alpha)\lambda_n + \alpha]\|\mathbf{s}\|^2 \end{aligned}$$

where λ_n is the largest eigenvalue of C. Since $I - C$ is positive definite (by the definition of a CAR process), $\lambda_n < 1$ and hence the largest eigenvalue of U is strictly less than one. Conversely,

$$\begin{aligned} \mathbf{s}^{\mathrm{T}}U\mathbf{s} &\geqslant (1-\alpha)\lambda_1\|\mathbf{s}\|^2 - \alpha\|H\mathbf{s}\|^2 \\ &\geqslant (1-\alpha)\lambda_1\|\mathbf{s}\|^2 - \alpha\|\mathbf{s}\|^2 \geqslant -[\alpha - (1-\alpha)\lambda_1]\|\mathbf{s}\|^2 \end{aligned}$$

where λ_1 is the smallest eigenvalue of C, since H is a (sub-) stochastic matrix. Thus the contraction property is proved if $\lambda_1 > -1$, which we have to establish for each C but is true in most of our examples.

Examples

The simplest possible choice for C is ϕN with ϕ just less than the reciprocal of r, the number of neighbours of a pixel. This causes difficulty at the edges, since

$$\mathrm{E}(S_i \mid S_j, j \neq i) = \sum_{j \neq 1} C_{ij} S_j$$

and the edge sites will have a smaller sum and so have conditional mean pulled towards zero. In the examples we use toroidal edge correction to avoid this problem. Fortunately, in astronomical images the edges are in regions of background luminosity and thus at a fairly constant level. One alternative is to set a one pixel wide border, smooth **Z** to find the values on that border and then apply (7a) to interior sites. Yet another suggestion (of Rafael Molina) is to take

$$\text{var } \mathbf{S} = \kappa_s (P - \phi N)^{-1}$$

$$P = \text{diag}\{(\#\text{nhbrs of } i)/r\}$$

when

$$\text{E}(S_i | S_j, j \neq i) = (r\phi) \times \text{average of } \mathbf{S} \text{ over nhbrs of } i$$

$$\text{var}(S_i | S_j, j \neq i) = \kappa_s / p_{ii}^2$$

Note that we cannot just replace ϕN by $r\phi\bar{N}$ where $\bar{N}$ averages over neighbours, since $\bar{N}$ is not symmetric.

Figure 5.3 shows a one-dimensional test image and reconstructions from noise with α chosen by eye. One-dimensional illustrations are easier to visualize, but the performance is similar both in two dimensions and with Poisson rather than Gaussian noise. Figure 5.4 shows reconstructions in the presence of blurring. Note that the value of α used can be larger in the presence of blurring since the H^{T} term in (7a) itself smooths **S**.

Figure 5.3 (a) A one-dimensional test image for reconstruction procedures. There are 101 pixels, with luminosities in the range (0, 1). (b) Added noise of variance $(0.02)^2$. (c) Reconstruction with $\alpha = 0.2$. (d) Reconstruction with $\alpha = 0.05$.

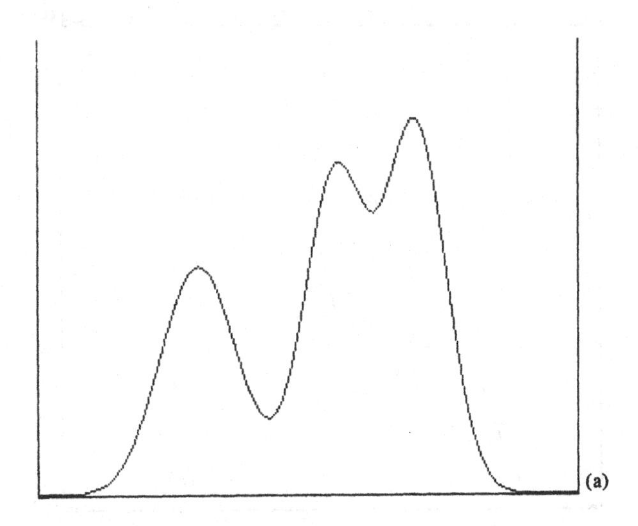

Figure 5.3 Continued.

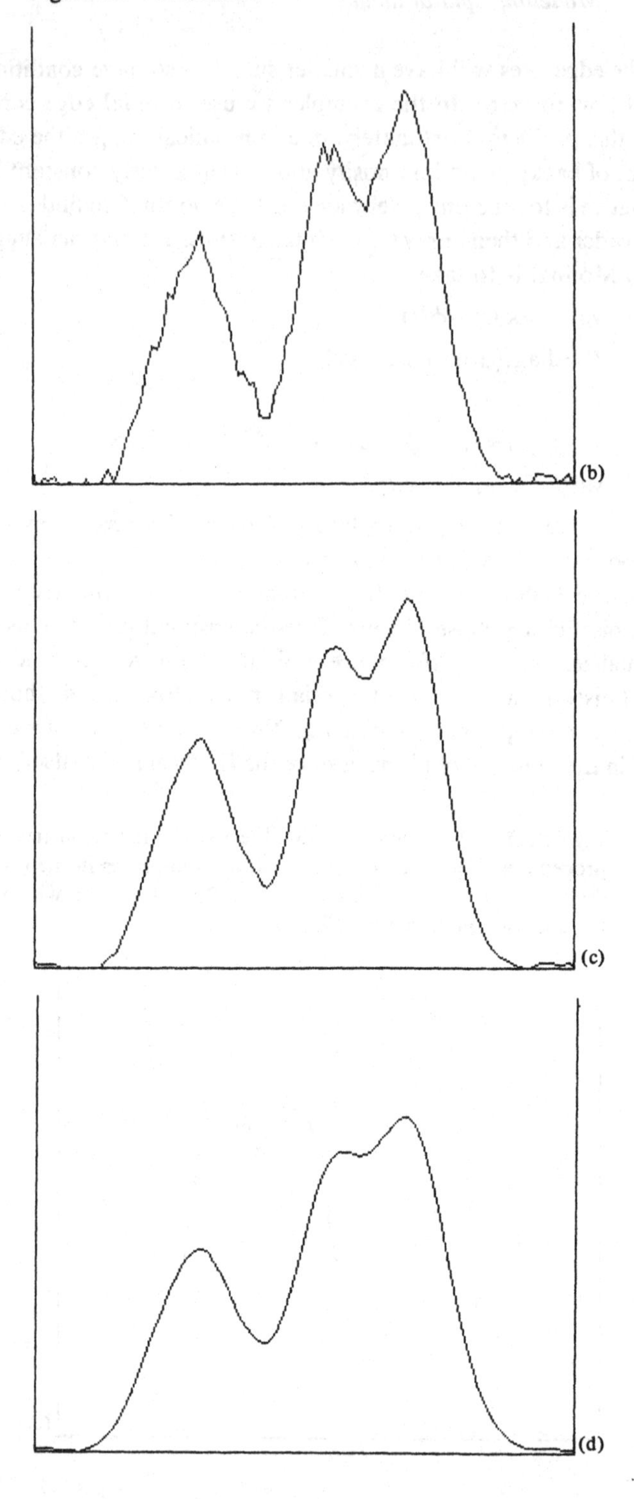

Figure 5.4 A test of deconvolution. (a) Blurred version of figure 5.3a with added noise. The blurring function is shown inset. (b) Deconvolution by (7a) with $\alpha = 0.4$, compared with true image.

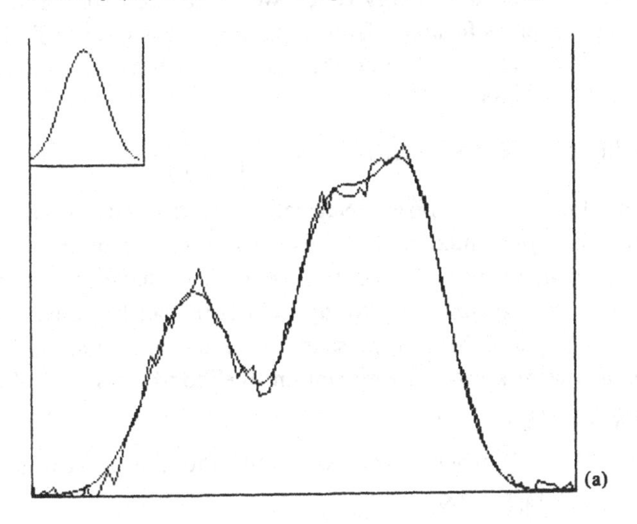

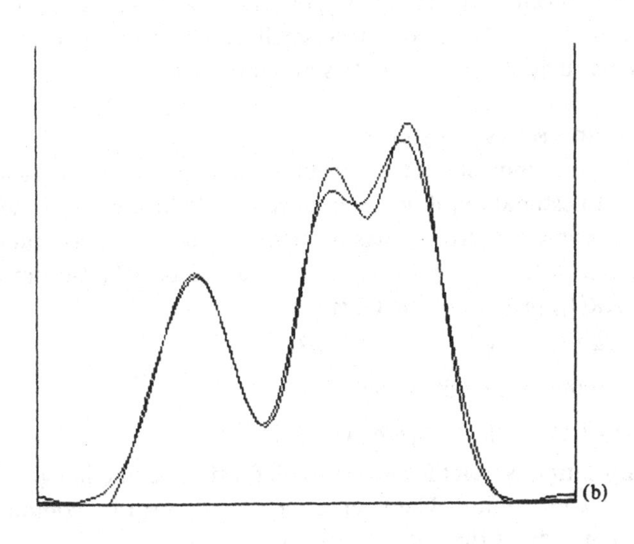

With the larger values of α convergence is rapid and 10–25 iterations suffice. With $\alpha \leqslant 0.05$ up to 100 iterations are needed. These values assume a sensible starting point for the iteration, for example $\mathbf{Z}$. Convergence occurs from any starting point but is much slower with $\hat{\mathbf{S}}^{(0)} = \mathbf{0}$, for example.

Parameter estimation

Our experiments depended on choosing α by eye. Since astronomical experiments are repeated many times, we could use experience to learn about α. However, is formal inference possible? It is easy to find reliable estimators of κ_n. Let $\bar{N}=N/r$ be the matrix which averages over neighbouring values. Then

$$\mathrm{E}[\|\mathbf{Z}-\bar{N}\mathbf{Z}\|^2|\mathbf{S}]=\|H(I-\bar{N})\mathbf{S}\|^2+\left(1+\frac{1}{r}\right)n\kappa_n$$

for observations from n pixels. Neglecting the first term (which will be small since $\mathbf{S}$ is presumed smooth) gives us an estimator of κ_n based on $\|\mathbf{Z}-\bar{N}\mathbf{Z}\|^2$, which we can correct for bias once we have an estimate of $\mathbf{S}$. The problem is to estimate κ_s. Many estimators can be constructed, but they differ wildly and fail to give 'sensible' values of α for smoothing. For example, if $\mathbf{S}$ were known the maximum likelihood estimator of κ_s is

$$\hat{\kappa}_s=n^{-1}\mathbf{S}^{\mathrm{T}}(I-C)\mathbf{S}$$

(Ripley, 1981, p. 90) whereas the pseudo-likelihood estimator is

$$\tilde{\kappa}_s=n^{-1}\|(I-C)\mathbf{S}\|^2$$

Both are unbiased when applied to realizations of the presumed prior model, but when applied to figure 5.3a yield $\hat{\kappa}_s=5.8\times10^{-4}$ and $\tilde{\kappa}_s=1.7\times10^{-5}$, too large and too small respectively. The smoothing experiments suggest $\kappa_s\approx1\times10^{-4}$ as an appropriate value.

A time series perspective

We can understand the difficulties inherent in an empirical Bayes procedure to estimate κ_s if we look more closely at the one-dimensional case. Then our CAR process has a one-sided representation (unlike the spatial case). In fact our prior is, up to end effects, a first-order autoregressive (AR(1)) process of the form

$$S_i=\beta S_{i-1}+\eta_i, \qquad \eta_i\sim N(0,\sigma^2)$$

with the parameter identification

$$\phi=\beta/(1+\beta^2), \qquad \kappa_s=\sigma^2/(1+\beta^2)$$

The identification is exact for processes defined on all the integers or with toroidal edge correction. The later is most easily seen by computing the covariance matrix of the AR process. Let

$$A=\begin{bmatrix} 1 & & & -\beta \\ -\beta & 1 & & \\ & & \cdots & \\ & & -\beta & 1 \end{bmatrix}$$

Then $A\mathbf{S}=\boldsymbol{\eta}$ and so $\mathbf{S}$ has covariance matrix $\sigma^2(A^{\mathrm{T}}A)^{-1}$. Since $A^{\mathrm{T}}A$ is a circulant matrix with row $(0, \ldots, 0, -\beta, 1+\beta^2, -\beta, 0, \ldots, 0)$ we can identify $\sigma^{-2}A^{\mathrm{T}}A$ with $\kappa_s^{-1}(I-\phi N)$.

We have taken ϕ to be just less than 0.5 and hence β just less than one. Our prior model for $\mathbf{S}$ is thus revealed as a random walk. Whereas figure 5.3a has the gross features of a random walk on a circle (so constrained to return to near zero at the last pixel) except that its excursions are always positive, its fine structure is too smooth. Both $\hat{\kappa}_s$ and $\tilde{\kappa}_s$ rely on the fine structure to estimate κ_s. This is obvious for the pseudo-likelihood estimator. For $\hat{\kappa}_s$ we can consider $\hat{\sigma}^2=n^{-1}\|\mathbf{S}-\beta B\mathbf{S}\|^2$, where B is the backshift matrix (on a circle). This is related to $\hat{\kappa}_s$ by $\hat{\kappa}_s=\hat{\sigma}^2/(1+\beta^2)$. From this point of view even $\hat{\sigma}^2\approx 1.17\times 10^{-3}$ is too low to explain the large changes of level in figure 5.3a. One idea is to use larger increments to estimate σ^2. For example

$$\mathrm{E}(S_i-S_{i-\tau})^2=\tau\sigma^2$$

and if we average such squared increments with τ about 10 we find $\hat{\sigma}^2\approx 6\times 10^{-3}$. We can even estimate in the presence of added noise since $\mathrm{E}(Z_i-Z_{i-\tau})^2=\tau\sigma^2+2\kappa_n$ and we have a reliable estimator of κ_n. However, such values of σ^2 imply that the small-scale structure of $\mathbf{S}$ is much rougher than we observe and so require too little smoothing to obtain $\hat{\mathbf{S}}$. We conclude that the AR(1)/random-walk model is unsuitable.

A frequency-domain analysis is also revealing. Consider n pixels. Let $U_{rs}=n^{-1/2}\exp(2\pi irs/n)$. Then because of the circular nature of our processes, U is the eigenvector matrix for all covariance matrices and also for N. Let * denote conjugation and transpose. Then for any stationary process $\Sigma=U\Lambda U^*$ where Λ is the diagonal matrix with entries $2\pi f(2\pi s/n)$. Further, (6) can be expressed as

$$(|h|^2+\lambda I-\lambda c)(U^*\hat{\mathbf{S}})=h^*(U^*\mathbf{Z})$$

where h and c are diagonal matrices of eigenvalues of H and C. (This suggests a fast method of solution using fast Fourier transform algorithms, at least when n is a power of 2.) In our applications H will be symmetric and hence h is real. There are then no phase changes, but the filter $\mathbf{Z}\rightarrow\hat{\mathbf{S}}$ has gain function $h/[h^2+\lambda(1-c)]=h/[h^2+\kappa_n/2\pi f]$ where f is the spectral density assumed for $\mathbf{S}$. Since there was no loss of generality in assuming a CAR process, this result is true for any stationary prior (on a circle).

Consider the case of no blurring. The gain function is then $f_S/(f_S+f_{\mathrm{noise}})$ which shows that the smoothing gained at high frequencies depends on the decay of f_S, which is implied by the model chosen and the parameters which are chosen to fit well at low frequencies. The crucial nature of the form of the prior thus comes across very clearly.

Figure 5.5 shows some nonparametric spectral density plots of **S** and **Z** for figure 5.3. The spectral density of the white noise is 6.4×10^{-5} and so dominates f_Z at higher frequencies. It is clear that after first differencing the **S** process is far from white noise, and this is also true after second differencing. Nevertheless, it appears that second differencing may be more appropriate.

A second-difference prior

An alternative to the CAR prior is a simultaneous autoregression of the form

$$(I - \phi N)\mathbf{S} = \boldsymbol{\varepsilon}, \qquad \boldsymbol{\varepsilon} \text{ white noise var } \kappa_d$$

This has variance matrix $\kappa_d(I - \phi N)^{-2}$ and so can be reexpressed as a CAR process with

$$\kappa_d^{-1}(I - 2\phi N + \phi^2 N^2) = \kappa_s^{-1}(I - C)$$

The problem here is that N^2 has diagonal terms equal to r, and C must have a zero diagonal. Thus

$$\kappa_d^{-1}(I - \phi N)^2 = \kappa_d^{-1}[(1 + r\phi^2)I - 2\phi N + \phi^2(N^2 - rI)]$$

from which

$$\kappa_d = (1 + r\phi^2)\kappa_s$$

$$C = \phi[2N - \phi(N^2 - rI)]/(1 + r\phi^2)$$

Again the time series perspective is helpful. This model corresponds to an AR(2) process which in the limit as $\phi \to 1/r$ becomes second differencing. In that case C gives weights $(-\frac{1}{6}, \frac{2}{3}, 0, \frac{2}{3}, -\frac{1}{6})$ to the first and second neighbours on the left and right. This is a case in which (7) is not convergent for small α since C has eigenvalues less than -1. We can however find a stable iteration by

$$\hat{\mathbf{S}}^{(i)} = \tfrac{1}{2}\hat{\mathbf{S}}^{(i-1)} + \tfrac{1}{2}[\alpha \mathbf{Z} + (I - \alpha)C\hat{\mathbf{S}}^{(i-1)}]$$

since the smallest eigenvalue exceeds -2. This was used for the experiments shown in figure 5.6 but converged slowly. Fourier methods would be more appropriate for serious applications. The smoothing is generally more successful. Further, the maximum likelihood estimate of κ_s from **S**, 1.1×10^{-5}, corresponds to a realistic value of α. The spectral density plots of figure 5.5d, e show that there is a hope of finding essentially the same estimate of κ_d from **Z** as from **S**, since the low frequency characteristics are the same. The log-likelihood of (κ_n, κ_d) is given by

$$-2L(\kappa_n, \kappa_d) = \sum_{0}^{n-1} \ln f_Z(2\pi s/n) + \Sigma f_Z(2\pi s/n)^{-1} |(U^*\mathbf{Z})_s|^2 \qquad (8)$$

which can be maximized over κ_d alone (using a simple $\hat{\kappa}_n$) or over both

Figure 5.5 Nonparametric spectral density plots on $\log_{10}$ scale. (a) Figure 5.3a. (b) First differences of figure 5.3a. (c) Figure 5.3b. (d) Second differences of figure 5.3a. (e) Second differences of figure 5.3b.

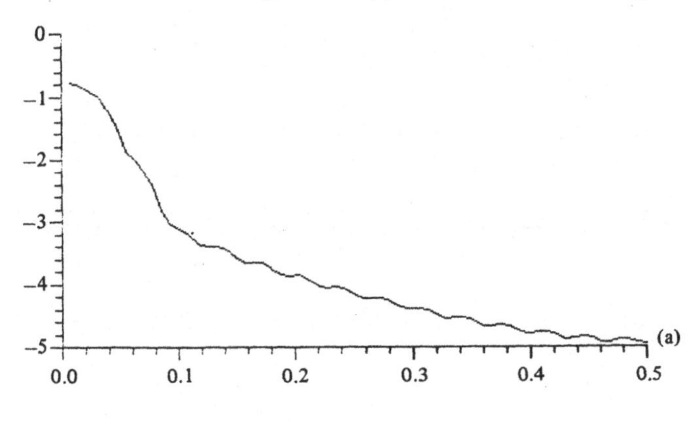

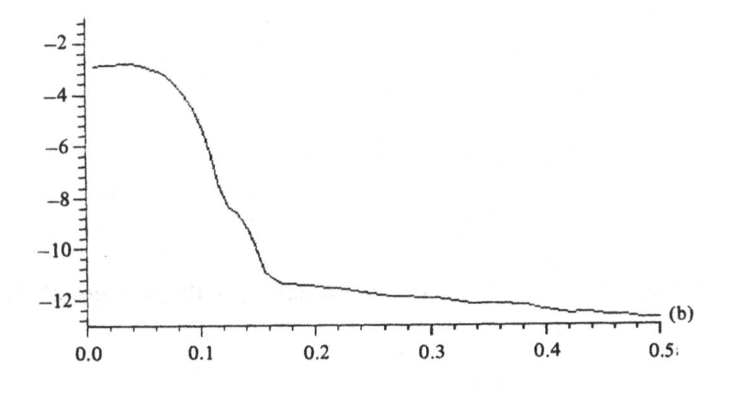

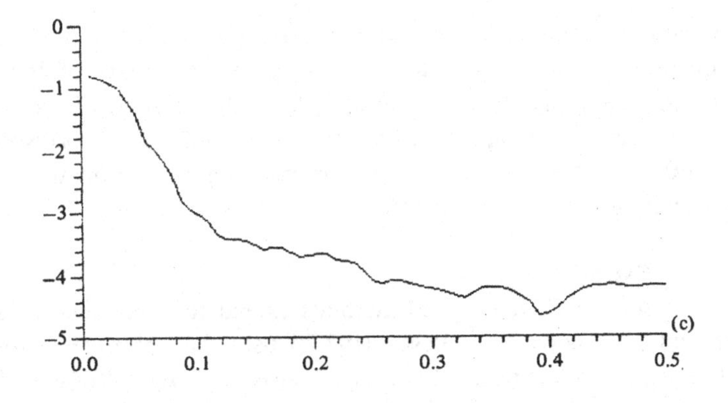

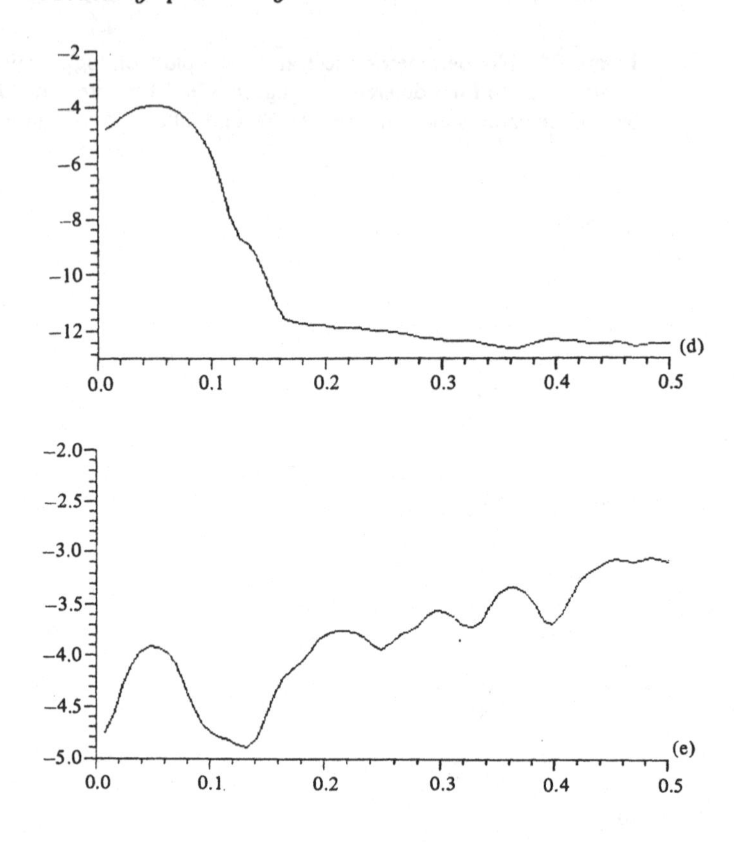

parameters. This provides a way to estimate κ_d in the presence of blurring, for

$$f_Z(\omega)=\frac{\kappa_n}{2\pi}+|h(\omega)|^2 f_S(\omega)$$

will contain information about the scale parameter κ_d in f_S at low frequencies. Maximizing (8) for the noisy unblurred image of figure 5.3b gave an almost unchanged value of $\hat{\kappa}_d$ from that obtained from **S**. With blurring there was an appreciable change to $\hat{\kappa}_d=6.3\times10^{-6}$ corresponding to $\alpha=0.016$. Figure 5.6c shows the corresponding restoration, which is less good but nevertheless acceptable.

Extensions

All the Fourier-based methods extend to two dimensions with toroidal edge correction. Besag (1977b) uses the appropriate tools to estimate parameters in f_S with additive noise, in an agricultural field trials context. In particular, fast Fourier transforms provide an efficient way to compute $\hat{\mathbf{S}}$ for 128^2 to 512^2 digital astronomical images.

Figure 5.6 Reconstructions with a second-difference prior. (a) Without blurring. (b) With the blur shown in fig. 5.4, compared with the true image. In both cases $\alpha=0.028$ corresponds to the maximum likelihood estimate of κ_d from **S**. (c) $\alpha=0.016$ corresponding to estimation from **Z**.

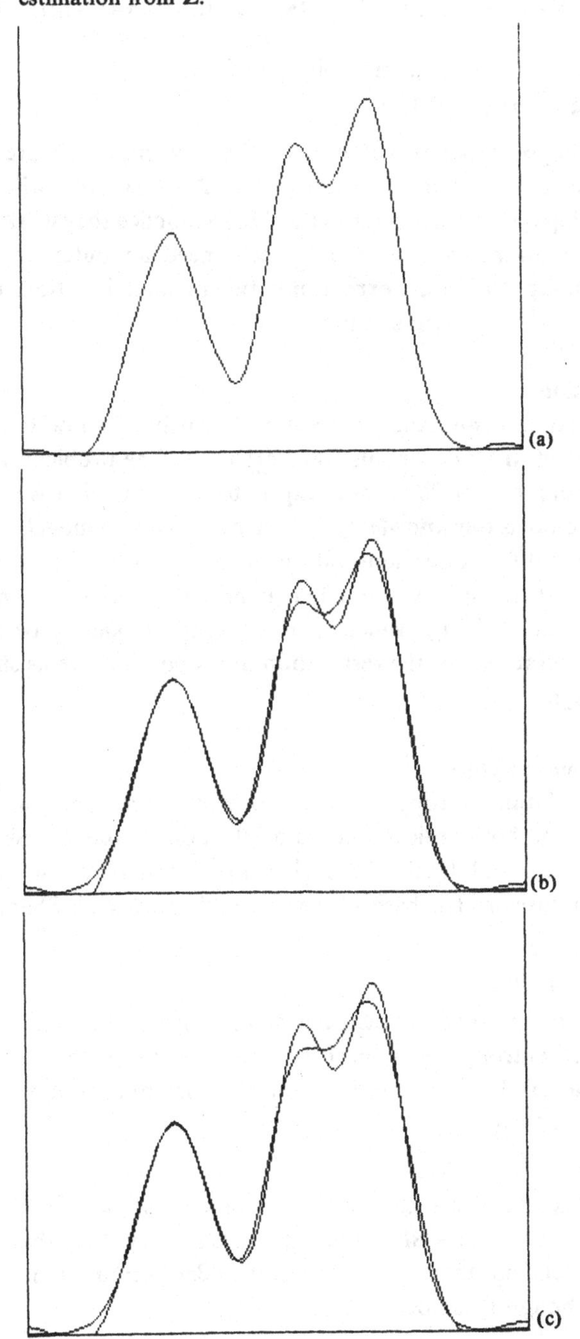

For Poisson noise the posterior density becomes

$$-2 \ln \mathrm{P}(\mathbf{S}|\mathbf{Z}) = \text{const.} + 2 \sum_i [(H\mathbf{S})_i - Z_i \ln(H\mathbf{S})_i] + \frac{1}{\kappa_s} \mathbf{S}^{\mathrm{T}}(I-C)\mathbf{S}$$

which has a stationary point at the solution of

$$(I-C)\mathbf{S} = \kappa_s H^{\mathrm{T}} D(\mathbf{Z} - H\mathbf{S})$$

where D is the diagonal matrix with $d_{ii} = 1/(H\mathbf{S})_i$. For this all the previous iterative schemes can be derived, although the $(\mathbf{Z} - H\mathbf{S})$ terms will have weights depending on the current estimate of **S**. In practice they work well. Fourier-based solutions can be used but will need an outer iteration updating D, although we would expect only two or three iterations to be needed provided $\mathbf{S}^{(0)}$ is chosen sensibly.

Discussion

These experiments (and others not shown here) show that the general principle of MAP deconvolution works well. The problems all lie in the lack of fit of the prior. We cannot expect to be able to choose a prior that accurately models our knowledge of images, and crude models prove to be less successful for continuous-valued images than for the k-valued images discussed in the next section. This is probably due to the 'noise' inherent in a k-valued image which allows a crude model to be used, whereas our expectations for the restoration of a smooth curve as shown here are very high.

Maximum entropy

The maximum entropy approach to inverse problems has been widely advocated in both optical and radio astronomy. Gull and Skilling (1985) and Skilling and Gull (1985) give an extensive review and a philosophical discussion. The basis of the 'maxent' approach is Shannon's entropy

$$\mathrm{SE} = -\Sigma p_i \log p_i$$

applied to a discrete probability distribution (p_i). Gull and Skilling argue that the correct entropy for a positive image is SE with $p_i = s_i/\Sigma s_j$ corresponding to pixel i. Their method is then to maximize **S** subject to a fidelity constraint such as

$$C = \Sigma(Z_{ij} - HS_{ij})^2/\sigma_{ij}^2 \leqslant C_{\mathrm{aim}}$$

We recognize this as a special case of the general discussions of §5.2, with C as the infidelity term and $-$SE as the roughness penalty. Note that $-$SE is position independent, that is it is invariant under permutations of the pixel values (although C is not).

The roughness $-SE$ is minimized by constant p_i, hence constant S_i. Thus the 'maximally noncommital' reconstructions claimed for maxent are in fact as uniformly grey as possible. Note that this is a global property, so reconstructing an image with two dissimilar areas will give different results whether treated as one or two images. Another example of the difference between our approach and maxent is a chequerboard pattern with two intensities close together. This would be smooth under $-SE$ but rough under the CAR model. Conversely, figure 5.3a is fairly smooth under the CAR prior but rough under $-SE$. It is the global nature of SE which causes the computational costs of using SE.

The arguments for SE are based on statistical physics and have an almost religious flavour. The language of statistical physics has hidden independence assumptions which lead to the permutation invariance of SE, and do not correspond to the structures we are seeking. Jaynes (1985, p. 398) proposed a modified entropy $-\Sigma p_i \log(p_i/m_i)$ which allows prior knowledge to be incorporated in $\mathbf{m}$. In general though, maxent and our smoothing have different aims and hence use different prior information.

5.4 APPLICATION TO SEGMENTATION

Segmentation has been the application most discussed by statisticians, and indeed was the application used by Geman and Geman (1984) in expounding the general Bayesian approach. The challenge of the segmentation problem is that (4) must be minimized over c^{MN} maps, a very large combinatorial optimization problem. The Gemans applied *simulated annealing*, an optimization technique popularized by Kirkpatrick, Gellatt and Vecchi (1983) but dating back to Pincus (1968, 1970).

The idea of simulated annealing is simple. Let P denote a probability distribution over a finite set $\mathscr{X}$. We wish to choose the member x^* of $\mathscr{X}$ with largest probability $P(x)$. Suppose we consider

$$P_\lambda(x) = P(x)^\lambda \Big/ \sum_{y \in \mathscr{X}} P(y)^\lambda$$

Then as $\lambda \to \infty$, P_λ increasingly concentrates on x^*. In particular, if we take a series of samples x_λ from P_λ as $\lambda \to \infty$, we would expect $x_\lambda \to x^*$ in some sense, preferably almost surely. Certainly sampling from P_λ with λ large many times and computing $P(x)$ in each case ought to turn up samples with near-maximal $P(x)$, and may well turn up x^*.

Suppose $P \propto \exp(-\beta U)$ is a Gibbs probability distribution. Then P_λ corresponds to replacing β by $\beta\lambda$. Since in statistical physics β is inversely proportional to temperature, it is usual now to write $\lambda = 1/T$ and talk about decreasing temperature to zero. Note that *any* function U will define a probability, so this technique is a general combinatorial minimization technique. We do need to be able to sample from P_λ. In real applications $\mathscr{X}$ will be far too large to be enumerated, so conventional methods of generating samples from a discrete distribution (Ripley, 1987, §3.3) are out of the question. As Pincus proposed, this is a natural problem for the iterative methods of Metropolis type (Metropolis *et al.*, 1953; Ripley, 1987, §4.7). These construct a Markov chain X_n with equilibrium distribution P. If we change P_λ and hence the transition matrix of the chain with time, we may get X_n to converge to x^*. This mimics the physical process of annealing an alloy, when it is cooled slowly so that the atoms can be arranged in a low-energy configuration.

Conditions on $\lambda(n)$ have been investigated by Geman and Geman (1984), Gidas (1985) and Mitra, Romeo and Sangiovanni-Vincentelli (1986). These authors show that $\lambda \sim \text{const.} \ln(n)$ is necessary for convergence *in probability* of X_n to x^*. This rate is so slow that it cannot be used. What we can do is to increase λ at about this rate and watch the value of (4), recording the map with the smallest value found. It may be useful to use coordinate descent (take $\lambda = \infty$) for the last few steps to obtain a low value of (4), and to run the algorithm several times and report the lowest minimum found. The combinatorial optimization literature suggests the use of a faster rate than $\ln(n)$ with this approach.

It is easiest to explain the implementation of the algorithm in a simple case. Consider the model (b) of §5.2 and additive Gaussian noise of variance κ. Then

$$P(\mathbf{l}) \propto \exp(\beta \times \text{number of neighbour pairs of the same label})$$

$$P(\mathbf{Z}|\mathbf{l}) = \exp\left[-\frac{1}{2\kappa} \sum_{\text{pixels}} (Z_{ij} - \mu_{l_{ij}})^2\right]$$

where μ_k is the luminance of type k pixels. Thus

$$\ln P(\mathbf{l}|\mathbf{Z}) = \text{const} - \frac{1}{2\kappa} \sum (Z_{ij} - \mu_{l_{ij}})^2 + \beta \sum_{\text{pairs}} 1(\text{same label}) \quad (9)$$

and the quantity to be minimized is

$$E = \Sigma(Z_{ij} - \mu_{l_{ij}})^2 - 2\beta\kappa \sum_{\text{pairs}} 1(\text{same label})$$

To sample from this posterior distribution we use the discrete analogue of the Ripley (1977) procedure, which Geman and Geman (1984) call the *Gibbs sampler*. For each pixel in turn we replace the current label l_{ij} by a sample from $P(l_{ij}|\mathbf{Z}$, other labels). From (9)

$$P(l_{ij}=k\,|\,\mathbf{Z}, \text{other labels})$$

$$\propto \exp\left[-\frac{1}{2\kappa}(Z_{ij}-\mu_k)^2+\beta\sum_{\text{nhbrs}} 1(\text{label } k)\right]$$

and we can sample directly from this discrete distribution.

A number of authors including Wohlberg and Pavlidis (1985) and Besag (1986) have commented on the excessive computational requirements of simulated annealing. This has not been the author's experience if it is implemented efficiently. The examples below on 64×64 binary images were computed in less than five minutes each on a Corvus Concept, a MC68000-based workstation. The programs used were entirely in Pascal and run twice as fast on an Atari 1040ST, a sub-£600 computer. In contrast, Wohlberg and Pavlidis quote 4 seconds per sweep (ca. 500 required) for a 32×32 image on a VAX 11/750, at least ten times more powerful a machine than the Corvus.

This perceived computational constraint led Besag (1983) to propose the use of coordinate ascent. At each pixel $P(l_{ij}=k\,|\,\mathbf{Z}$, other labels) is maximized. This corresponds to $\lambda=\infty$ in the Gibbs sampler if the pixels are updated sequentially. An alternative is to update the pixels synchronously, so a new label is computed for each pixel using the old value of the neighbouring pixels, and then all old labels are replaced by the new ones. This process is continued for a fixed time or until no further changes occur. The sequential version is guaranteed to move downhill and so will stop, but the synchronous version can oscillate. Besag calls this method ICM. A further refinement is to start with $\beta=0$ and increase it to the target value during the iteration. For the 64×64 test images below ICM took about 40 seconds. Although originally proposed as an approximation to minimizing E, Besag (1986) now regards ICM as desirable in its own right.

Parameters

Some simple geometric arguments allow us to guess values for β in simple cases. Suppose that there are c labels and 8 neighbours. In a situation such as

$$\begin{matrix} A & A & A \\ A & ? & A \\ A & A & A \end{matrix}$$

our prior opinion that $?=A$ must be high. Since

$$P(?\neq A|\text{other labels}) = 1-\frac{e^{8\beta}}{e^{8\beta}+(c-1)} = \frac{(c-1)}{(c-1)+e^{8\beta}}$$

must be small, less than 0.1% say, we find a lower bound on β. On the other hand we must allow sharp corners. Consider

$$\begin{matrix} A & A & A \\ A & ? & B \\ A & B & B \end{matrix}$$

Then

$$P(?=B|\text{other labels})=\frac{e^{3\beta}}{e^{3\beta}+e^{5\beta}+(c-2)}$$

must not be too small, say at least 10%. In the case $c=2$ this gives bounds $0.86<\beta<1.10$.

In the case of 4 neighbours we have

$$\begin{matrix} & A & \\ A & ? & A \\ & A & \end{matrix}$$

and

$$P(?\neq A|\text{other labels})=\frac{(c-1)}{(c-1)+e^{4\beta}}$$

again establishing a lower bound on β, this time twice that of the previous case. Sharp corners are now much more plausible, for in the case

$$\begin{matrix} & A & \\ A & ? & B \\ & B & \end{matrix}$$

$?=A$ and $?=B$ are equally likely (and somewhat more likely than any other label). This greater acceptability of sharp corners is clear from figures 5.14 and 5.15 below. In general we expect β to be about twice as large for a 4-neighbour graph as for an 8-neighbour one.

In most cases we will be able to fix parameters in the prior model by eliciting our prior opinions. In the process of doing so it is often helpful to look at simulations of the prior. As we have already seen, simulations of the prior itself can be often all of one label and so uninformative. Simulation experiments on the restoration of a test pattern can be extremely helpful. The single-label realizations of the model only apply with no externally imposed variation (a magnetic field in one physical context of the Ising model) and so are relevant to P(I) but not to P(I|Z).

The parameters in the infidelity I_0 may also be known from past experience, and can also be estimated from the restoration together with any remaining parameters in the prior. We then have an empirical Bayes procedure. Too few experiments have been done to date to comment on the efficacy of these procedures.

Examples

All our examples are binary, with two classes represented by black and white. This is partly for ease of reproduction but also because no additional features emerge for more than two colours.

We do achieve some computational simplification in the binary case. We find

$$\frac{\mathrm{P}_\lambda(l_{ij}=\text{black}|\mathbf{Z}, \text{rest of } \mathbf{l})}{\mathrm{P}_\lambda(l_{ij}=\text{white}|\mathbf{Z}, \text{rest of } \mathbf{l})}=\exp\left[\frac{\lambda}{\kappa}(Z_{ij}-\nu)+\beta\lambda(B-W)\right] \quad (10)$$

where $\nu=\frac{1}{2}(\mu_{\text{black}}+\mu_{\text{white}})$ and B and W are the number of black and white neighbours respectively. Besag's rule corresponds to $\lambda=\infty$ and so chooses black if and only if

$$Z_{ij}>\nu+\beta\kappa(B-W)$$

The simplicity of (10) illustrates how simulated annealing can be implemented efficiently. For both annealing and ICM it is worth recording $B-W$ for each pixel and updating it when a neighbour changes colour. For most pixels the odds ratio in (10) is going to be rather large, especially late on in the iterative procedure when λ is larger than one. It is not worth sampling l_{ij} on each scan of the pixels when most sites will be unchanged. Instead we can set a 'clock' at each pixel giving the time to the next change. This clock will be reset when a neighbour changes colour and so alters $B-W$. When the temperature changes at every sweep the computation of the time to next change is complicated. When temperature is constant, the time is geometrically distributed and so easy to generate. This is useful for annealing strategies in which the temperature is lowered in steps, and for the MPM runs in §5.5.1 in which temp$\equiv$1. In general, it can be used to provide a lower bound on the time to the next change.

Some housekeeping will be saved by setting the clock to $+\infty$ when the odds ratio is large, say 1000:1 or more, at the cost of a minor departure from the algorithm. (This is the active/inactive strategy of Ripley, 1986b.) Some experiments show that even more approximate strategies, concentrating only on sites with odds ratios of 10:1 or less, work just as well viewed as optimization algorithms. It is often best to do several runs with different random number seeds and take the best solution found as the answer. With these refinements simulated annealing runs very fast. Most of the ideas apply to c colour segmentation equally well but need more storage space.

In the particular case of binary images exact maximization algorithms can be used. Greig, Porteous and Seheult (in the discussion of Besag, 1986) point out that minimizing $-\ln \mathrm{P}(\mathbf{l}|\mathbf{Z})$ can be reformulated as a network flow problem and so solved exactly by the Ford-Fulkerson algorithm, albeit at vastly greater cost than the methods used here. It is also possible to use heuristics from combinatorial optimization algorithms, especially the graph partitioning problem. These are again approximate minimization techniques which compete successfully with simulated annealing on specific problems. On the other hand, annealing is a powerful general purpose technique.

Our first example (figures 5.7 to 5.11) was designed to be very different from realizations of the assumed prior (5). Figure 5.7 shows the pattern, and figures 5.8 to 5.10 restorations with three levels of noise. How do we

Figure 5.7 Test pattern for segmentation. (Figures 5.7 to 5.12 are reproduced with permission from Ripley, 1986b).

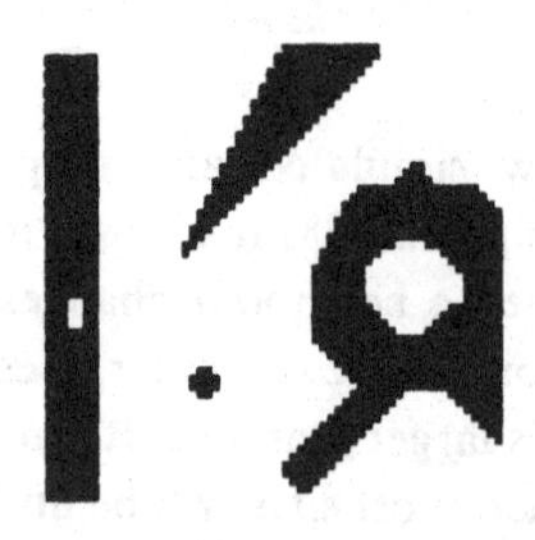

Figure 5.8 Segmentations from noise variance 0.2, $\mu_{\text{white}}=0$ and $\mu_{\text{black}}=1$. (a) Nonspatial, error rate 14.3%. (b) ICM, error rate 1.25%. (c) Simulated annealing, error rate 1.44%.

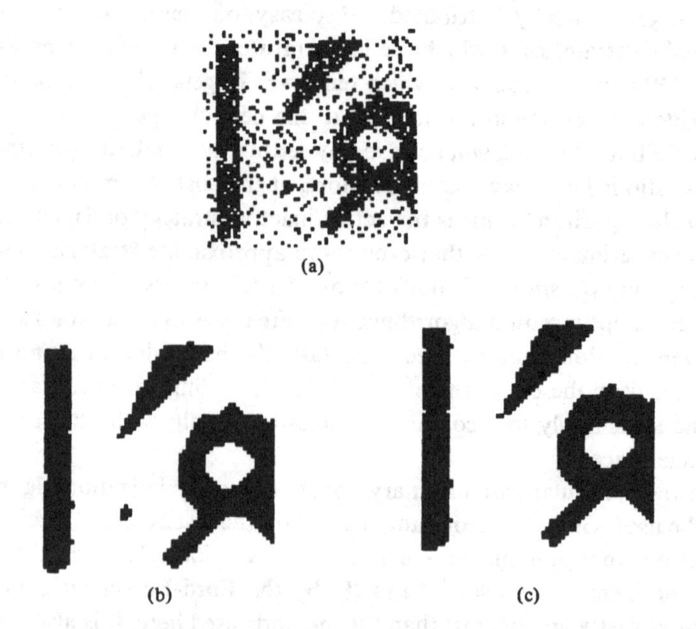

Figure 5.9 Segmentation with noise variance 0.5. Error rates are (a) 24.2%, (b) 4.08%, (c) 4.37%.

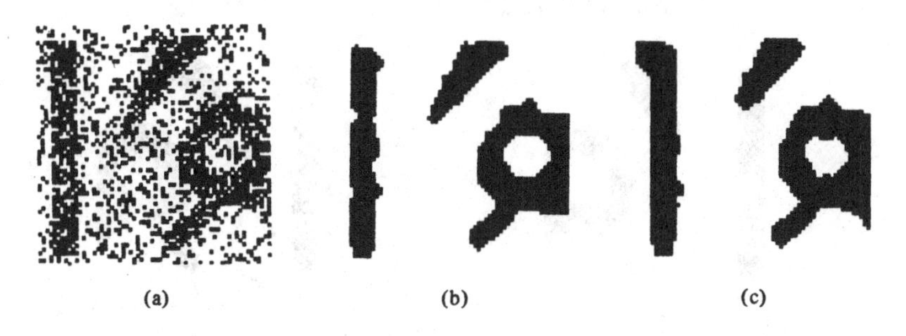

Figure 5.10 Segmentation with noise variance 1.0. Error rates are (a) 31.3%, (b) 9.23%, (c) 14.55%.

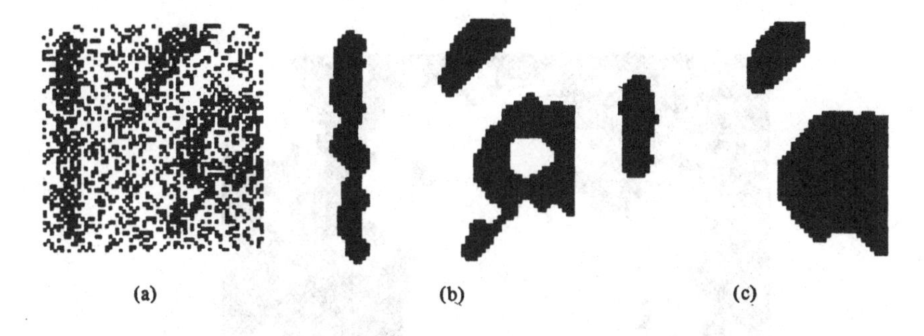

Figure 5.11 Two restorations with error rate 3.5%.

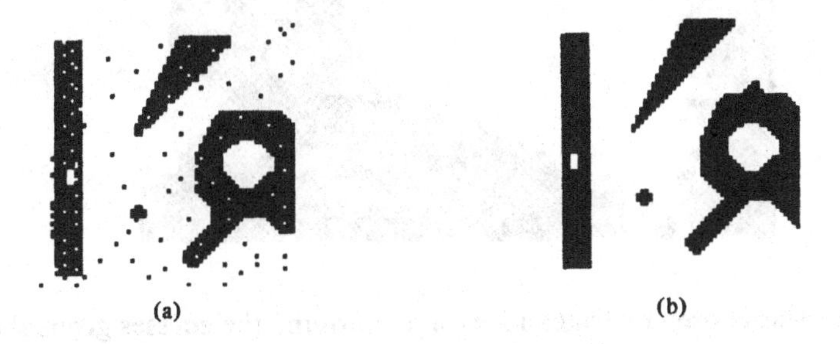

assess the performance of these segmentations? A traditional measure is the proportion of pixels which are misclassified, and this is shown in the captions. Figure 5.11 shows how misleading this can be. Both images have the same error rate but visually figure 5.11b is much more acceptable. The ICM results were done with $\beta=0.5$, 0.7, 0.9, 1.1, 1.3 and 1.5 ($\times 3$) with sequential updates (on Besag's recommendation). For global optimization

Figure 5.12 ICM segmentations of figure 5.7 with misspecified noise variances. (a) Actual $= 0.2$, $\kappa = 1.0$, error rate 7.89%. (b) Actual $= 1.0$, $\kappa = 0.2$, error rate 2.71%.

Figure 5.13 A test image, 64 × 64 pixels digitized from a map of Great Britain and Ireland. (Reproduced with permission from Ripley and Taylor, 1987.)

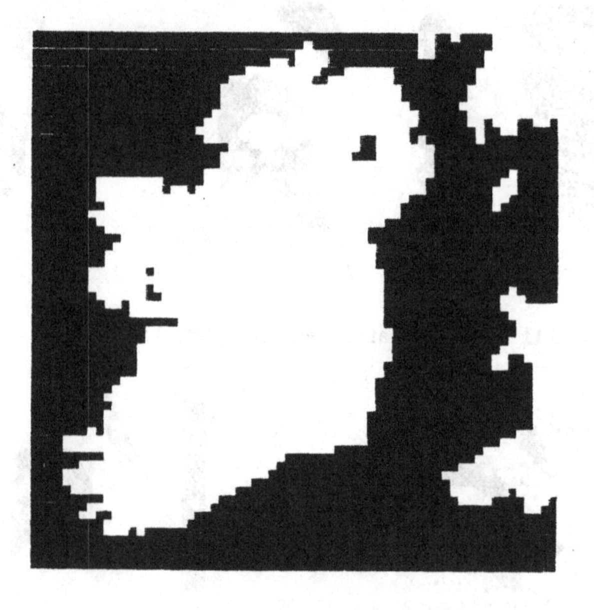

$\beta = 1$ was chosen by trial and error, confirming the guesses given above. In both cases 8 neighbours were used. Figure 5.12 shows the effect of misspecifying $\beta\kappa$.

Our second example (from Ripley and Taylor, 1987) is closer to our assumed prior. Again additive independent Gaussian noise was used. Figures 5.14 and 5.15 shows the effect of the choice of 4 or 8 neighbours and sequential or synchronous updates in ICM. In this and most other examples ICM works much less well with 4 neighbours rather than 8; in

Figure 5.14 Segmentations of figure 5.13 with additive noise, $\sigma = 0.65$. (a) sequential ICM, (b) synchronous ICM and (c) annealing with $\beta = 1$. In all cases 8 neighbours were used.

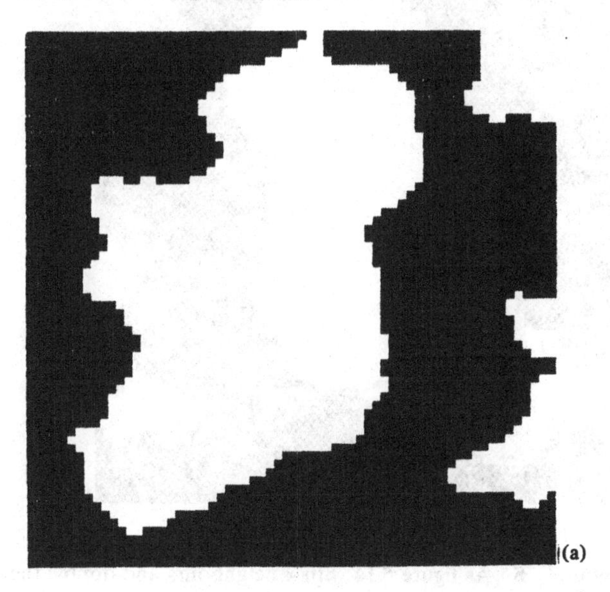
(a)

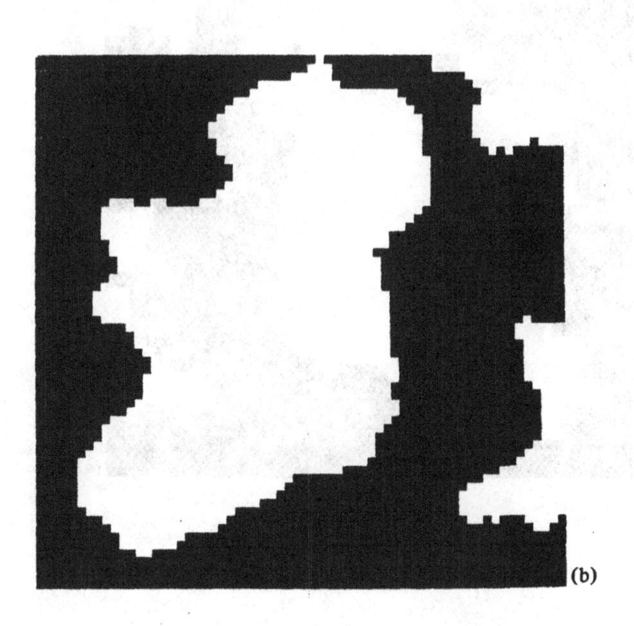
(b)

Figure 5.14 continued

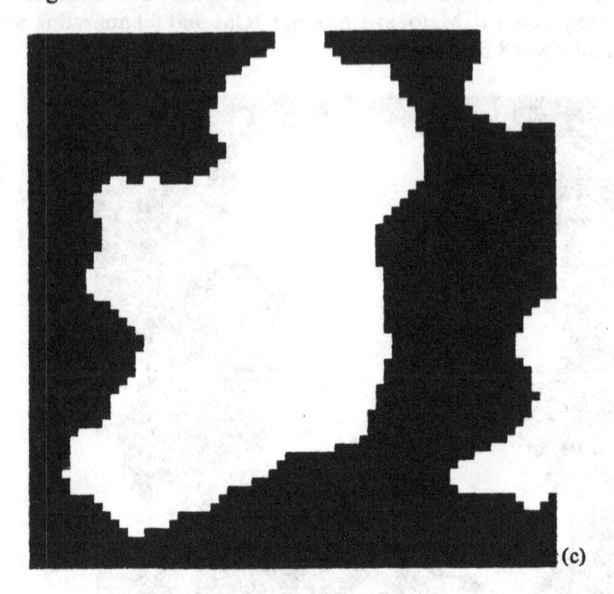

Figure 5.15 As figure 5.14 with 4 neighbours and double the values of β. The synchronous ICM solution oscillates.

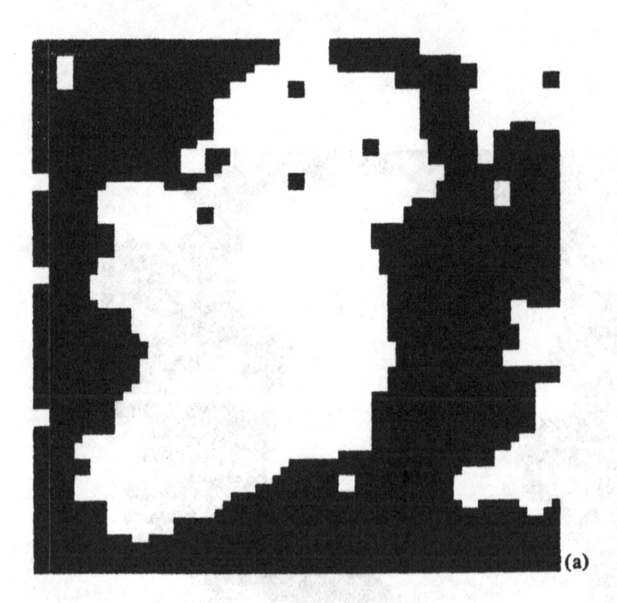

Figure 5.15 continued

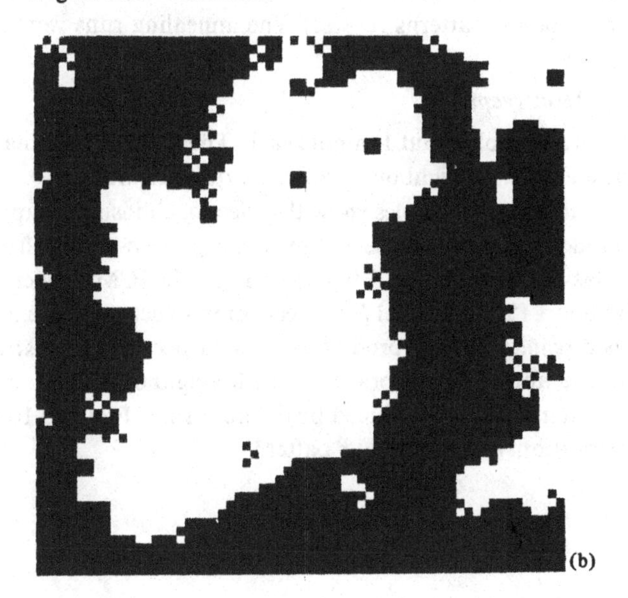

(b)

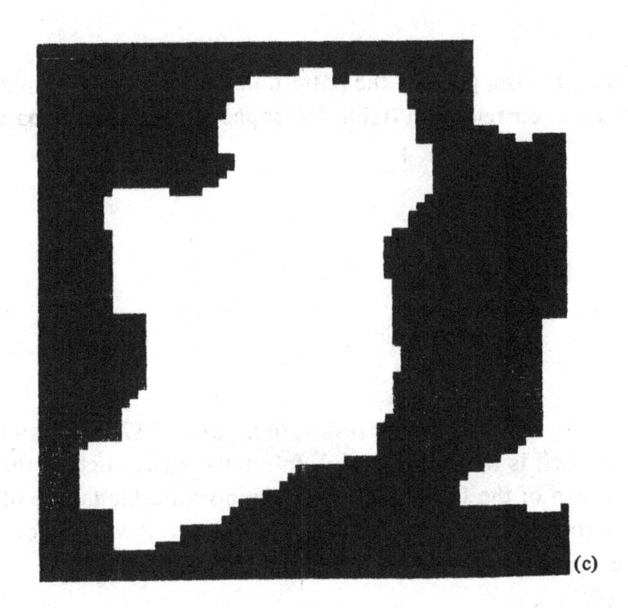

(c)

this case the synchronous ICM with 4 neighbours oscillates between two maps (the chessboard patterns reverse). The annealing runs were started from $\lambda=\frac{1}{2}$ with

$$\lambda=\tfrac{1}{2}\ln(nsweep)$$

for 200 sweeps and took about 1 minute each. The more angular nature of the restorations with 4 neighbours shows up quite clearly.

Our third and fourth examples show the effect of choosing an appropriate prior model. In neither case is any clear pattern visible from the nonspatial classification (figures 5.16a and 5.17a). The ICM segmentations with 8 neighbours (5.16b and 5.17b) reveal some structure. If we are told that there is a *regular* pattern present we may suspect diagonal stripes in figure 5.16. The annealing restoration with 8 neighbours is typical of a number of similar patterns produced by different runs. Figure 5.16 shows an ICM restoration with neighbour pattern

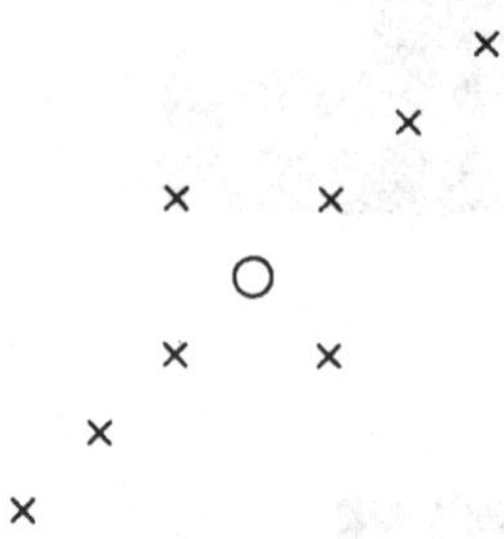

No observer has yet guessed the pattern in figure 5.17a. Looking at the two-dimensional correlogram (table 5.2) suggests a neighbour pattern

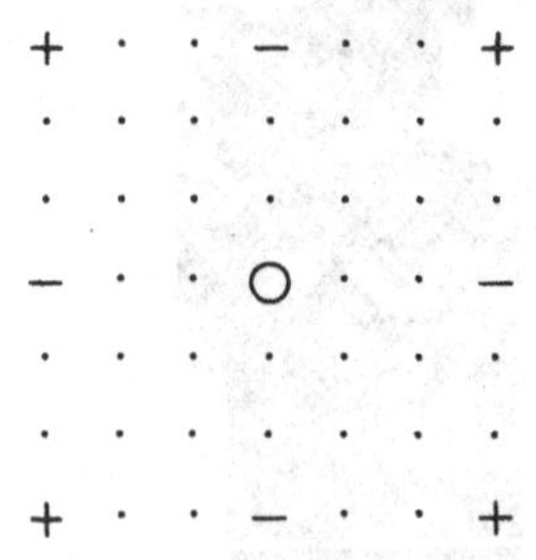

Figure 5.17d shows the ICM restoration, and 5.17e the annealing restoration (which is the true pattern). In regular cases such as these the order of the scan of the ICM procedure is important; the failure of 5.15d corresponds to the TV scan used, which limits the flow of information back to the upper left corner.

Evaluation

A major difficulty in performance evaluation of segmentation routines is deciding on a performance criterion. We have already seen the

Figure 5.16 A test for the choice of neighbours. Noise with $\sigma = 1$ was added. (a) Nonspatial segmentation. (b) ICM with 8 neighbours. (c) Simulated annealing with 8 neighbours. (d) ICM with another set of neighbours (see text). (e) True pattern.

(a)

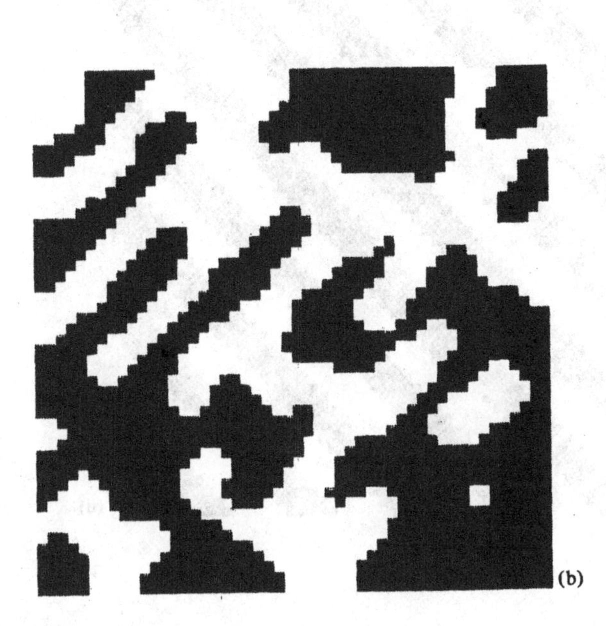
(b)

Figure 5.16 continued

(c)

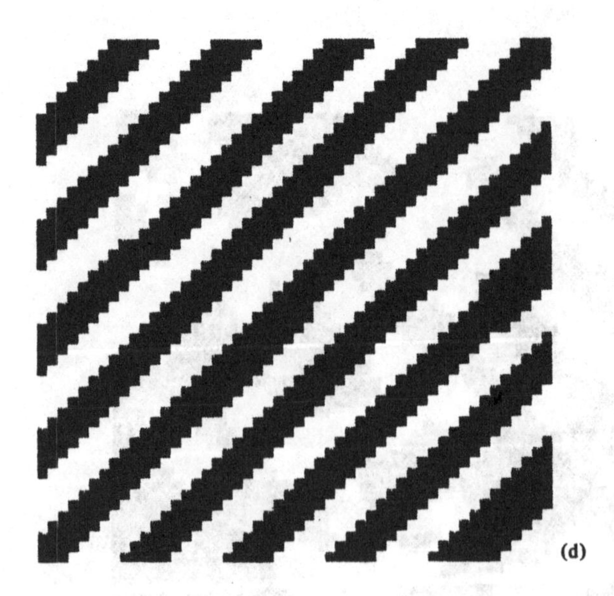

(d)

Figure 5.16 continued

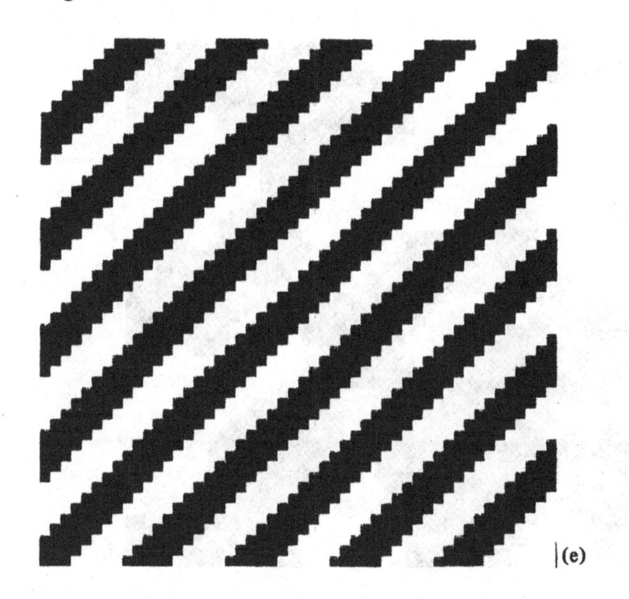

Figure 5.17 Another test. Details as figure 5.16.

Figure 5.17 continued

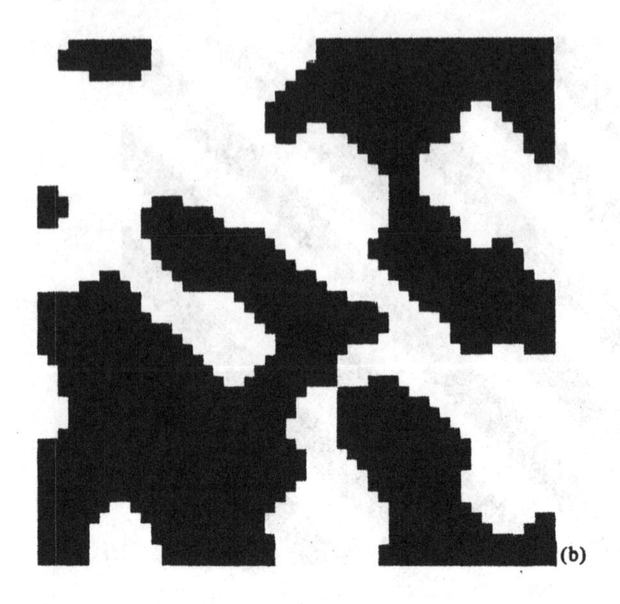

Figure 5.17 continued

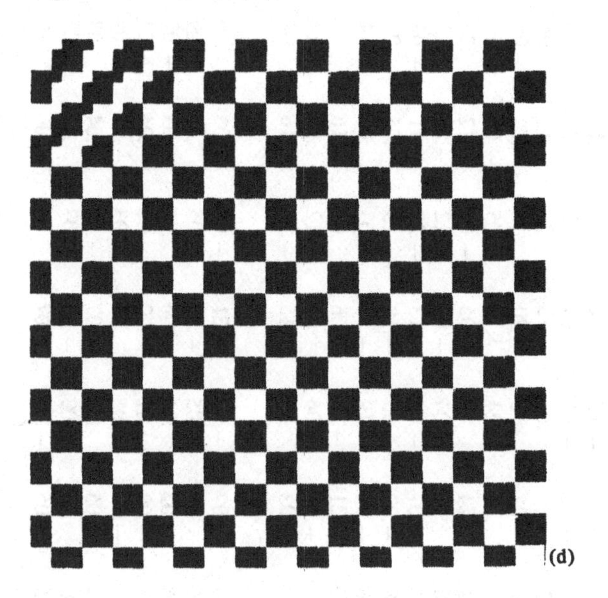

(d)

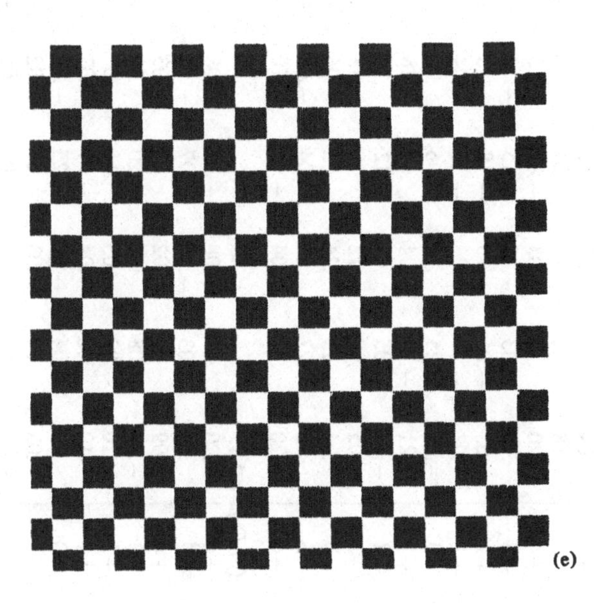

(e)

Table 5.2. *Correlograms for the signals for two restoration problems*

	−6	−5	−4	−3	−2	−1	0	+1	+2	+3	+4	+5	+6
(*a*) *Figure 5.16*													
+6	.04	.16	.21	.15	.03	−.03	−.16	−.22	−.16	−.04	.04	.12	.20
+5	.18	.22	.12	.04	−.06	−.14	−.23	−.16	−.06	.04	.12	.22	.15
+4	.20	.16	.03	−.06	−.13	−.21	−.12	−.03	.06	.13	.24	.13	.04
+3	.14	.03	−.03	−.15	−.19	−.12	−.06	.03	.13	.20	.12	.08	−.05
+2	0	−.05	−.13	−.18	−.15	−.02	.08	.12	.24	.12	.05	−.05	−.12
+1	−.03	−.11	−.23	−.13	−.04	.07	.11	.24	.18	.05	−.04	−.12	.21
0	−.13	−.25	−.12	−.06	.04	.14	1.00	.14	.04	−.06	−.12	−.25	−.13
−1	−.21	−.12	−.04	.05	.18	.24	.11	.07	−.04	−.13	−.23	−.11	−.03
−2	−.12	−.05	.05	.12	.24	.12	.08	−.02	−.15	−.18	−.13	−.05	0
−3	−.05	.08	.12	.20	.13	.03	−.06	−.12	−.19	−.15	−.03	.03	.14
−4	.04	.13	.24	.13	.06	−.03	−.12	−.21	−.13	−.06	.03	.16	.20
−5	.15	.22	.12	.04	−.06	−.16	−.23	−.14	−.06	.04	.12	.22	.18
−6	.20	.12	.04	−.04	−.16	−.22	−.16	−.03	.03	.15	.21	.16	.04
(*b*) *Figure 5.17*													
+6	.20	.09	−.08	−.19	−.08	.09	.20	.07	−.11	−.21	−.08	.06	.20
+5	.11	.03	−.04	−.07	−.04	.03	.06	0	−.05	−.08	−.05	.04	.09
+4	−.09	0	.02	.06	.03	−.02	−.06	−.02	.04	.06	.04	−.03	−.07
+3	−.21	−.08	.10	.20	.10	−.06	−.23	−.09	.07	.20	.06	−.04	−.23
+2	−.10	−.02	.05	.11	0	−.01	−.04	−.04	.05	.06	.03	−.04	−.07
+1	.10	.06	−.04	−.07	−.03	.05	.05	.05	.02	−.07	−.03	.04	.08
0	.22	.04	−.07	−.25	−.09	.09	1.00	.09	−.09	−.25	−.07	.04	.22
−1	.08	.04	−.03	−.07	.02	.05	.05	.05	−.03	−.07	−.04	.06	.10
−2	−.07	−.04	.03	.06	.05	−.04	−.04	−.01	0	.11	.05	−.02	−.10
−3	−.23	−.04	.06	.20	.07	−.09	−.23	−.06	.10	.20	.10	−.08	−.21
−4	−.07	−.03	.04	.06	.04	−.02	−.06	−.02	.03	.06	.02	0	−.09
−5	.09	.04	−.05	−.08	−.05	0	.06	.03	−.04	−.07	−.04	.03	.11
−6	.20	.06	−.08	−.21	−.11	.07	.20	.09	−.08	−.19	−.08	.09	.20

inadequacy of the misclassification rate, which indeed prefers figure 5.12b to 5.10b. Owen (1986) considers misclassifications for different classes of pixels (boundary/interior, label type, number of 'like' neighbours) but in his and other examples these criteria do not seem very informative. They would of course differentiate between figures 5.11a and 5.11b beyond any shadow of doubt. With more than two classes attention will need to be paid to what error is made, and one can produce a 'confusion matrix' giving the number of type i pixels classified as type j. This will highlight another failing of the misclassification rate, which ignores the possibility that errors in rare classes may be more important than those in common ones.

Performance evaluation is one of the most important research topics remaining in segmentation.

A forestry problem†

Commerical forests are often assessed in 'stands', relatively homogeneous subareas of about a prespecified area (which differs from region to region). The stands are the basis for assessing the value of the forest. A problem is to define them efficiently, and it has been suggested that this can be done from satellite images of the forest, picking stands which are homogeneous with respect to 'colour' (which includes infra-red information).

We can incorporate this problem within our general framework at (4). It *is* segmentation, but of a different type to those problems discussed so far in this section. The *ad hoc* approach is to select a fixed number of *seed pixels* whose neighbourhoods (3×3, say) have the lowest variances of all pixels, and to 'grow' strands from these seed pixels. In our framework, the infidelity term will reflect the inhomogeneity of a stand, and the energy the departure from our prior assumption that stands are roughly of prespecified size with smooth boundaries. This vague description would need to be converted to a mathematical model and simulated annealing could then be used to minimize (4).

5.5 OTHER STATISTICAL APPROACHES TO SEGMENTATION

The Bayesian approach considered in §5.2 and §5.4 which involves estimating the whole map **l** simultaneously is but one of a number of statistical approaches which have been proposed in the last five years. Most of these are based on estimating l_{ij} separately for each pixel (i, j).

† I am grateful to Dr Erkki Tomppo of the Technical Research Centre, Espoo, Finland, for bringing this problem to my attention.

5.5.1 Pixel-by-pixel Bayesian methods

The loss function corresponding to MAP estimation is 1 unless the whole map $\mathbf{l}$ is correct, otherwise 0. A commonly used measure of the error of a segmentation is the misclassification rate, corresponding to loss function the number of incorrect pixels. For this loss, the Bayes rule maximizes

$$\sum_{\text{pixels}} P(L_{ij}=l_{ij} \mid \mathbf{Z})$$

so each label l_{ij} is chosen as the marginal posterior mode of L_{ij}. This is sometimes also known as MAP estimation, but perhaps better as MPM estimation. Some authors (e.g. Derin *et al.*, 1984) regard MPM as an approximation to MAP estimation, whereas others (Marroquin, Mitter and Poggio, 1987) regard it as desirable in its own right. It can also be used for initial values for annealing.

Following standard practice in pattern recognition (e.g. Devijver and Kittler, 1982) we may allow a further 'don't know' decision D with loss $1/c<\alpha<1$ (whatever the true label). The Bayes rule then chooses

$$l_{ij}=k \quad \text{if } P(l_{ij}=k \mid \mathbf{Z})>1-\alpha$$
$$D \quad \text{if } \max P(l_{ij}=k \mid \mathbf{Z}) \leqslant 1-\alpha$$

For $\alpha \geqslant 1-\frac{1}{c}$ this reverts to the MPM rule.

The posterior marginal distributions do not have a simple form, and many of the methods given in the rest of this chapter seek to overcome this problem. There is a direct simulation-based way to find MPM estimates, hinted at by Grenander (1983) and used by Marroquin *et al.* (1987). If we simulate a series of observations from $P(\mathbf{l} \mid \mathbf{Z})$ (using the Gibbs sampler or another iterative sampling scheme) we can estimate the marginal distribution of $l_{ij}=k$ by its frequency in the series of runs. To compute figure 5.18 we ran the Gibbs sampler for 1000 runs and plotted the label for each pixel with the largest frequency in the series.

The problem with analytical attempts to find $P(l_{ij}=k \mid \mathbf{Z})$ is the summation of $P(\mathbf{l} \mid \mathbf{Z})$ over the labels at the other pixels. This can be reduced by approximating $P(L_{ij}=k \mid \mathbf{Z})$ by $P(l_{ij}=k \mid \mathbf{Z}_N)$ where $\mathbf{Z}_N$ denotes observations in a neighbourhood of (i, j). Several methods have been based on this observation. In the case of the chess and stripe patterns the approximation will be poor, but the approximate methods will be 'local' in character.

Swain, Vardeman and Tilton (1981) estimate $P(\mathbf{l}_N)$ directly from a training set and hence find $P(l_{ij}=k \mid \mathbf{Z}_N)$. Kittler and Foglein (1984) propose a number of approximations. The remaining approaches (below) model $\mathbf{L}_N$.

Figure 5.18 MPM segmentations corresponding to (a) figure 5.14, (b) figure 5.16 and (c) figure 5.17. In the last two cases the results varied considerably from run to run.

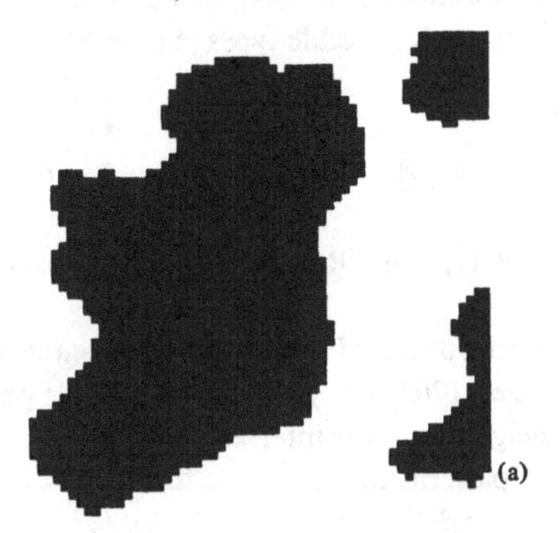

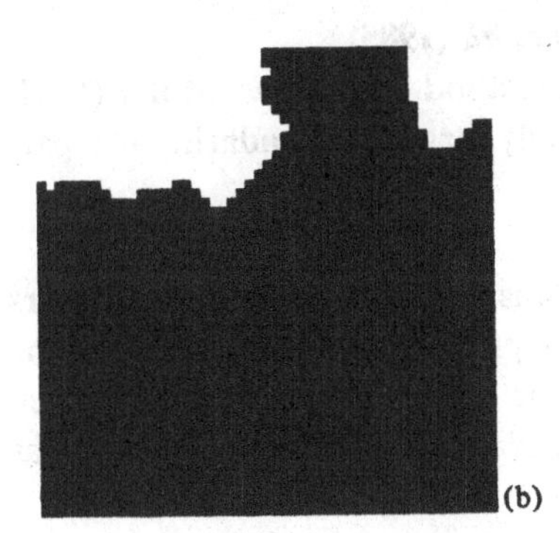

5.5.2 Owen (1984)

Owen used the approximate pixel-by-pixel method without a 'don't know' option and N defined by the pixel $C=(i,j)$ and its four horizontal and vertical neighbours (N, E, S, W). Then

$$\begin{aligned} P(L_{ij}=k \mid \mathbf{Z}_N) &= \sum_n \sum_e \sum_s \sum_w P(L_c=k, L_N=n, L_E=e, L_S=s, L_w=w \mid \mathbf{Z}) \\ &= \sum_{n,e,s,w} P(Z_C, Z_N, Z_E, Z_W \mid L_C, L_N, L_E, L_S, L_W) \\ &\quad \times P(L_N=n, L_E=e, L_S=s, L_W=w \mid L_C=k) \\ &\quad \times P(L_C=k) \qquad \textbf{(11)} \end{aligned}$$

The first term in the product is known if the noise is independent from pixel to pixel. The final term is the prior probability of class k, which is commonly taken to be $1/c$. The second term needs a model for $\mathbf{l}$. Owen assumed that the true map is a mosaic of polygonal regions of constant label. When this is digitized, the only possible types of pattern for the cross of five pixels are

$$\begin{array}{ccc} & A & \\ A & A & A \\ & A & \end{array} \qquad \begin{array}{ccc} & A & \\ A & A & B \\ & A & \end{array} \qquad \begin{array}{ccc} & A & \\ A & A & B \\ & B & \end{array}$$

given the central pixel is of type A. (Rotations of these patterns are allowed.)

Owen computed the probabilities of these types of patterns for a Poisson mosaic model (Switzer, 1965). They depend on a single parameter β, the probability that the neighbour cross intersects a boundary. Then the probabilities of the types of patterns are

$$1-\beta \qquad \alpha\beta/4 \qquad (1-\alpha)\beta/4$$

where $\alpha=\sqrt{2}-1$. Owen proposes the estimation of β from a training set containing crosses of known ground truth.

5.5.3 Hjort and Mohn (1984), Saebø *et al.* (1985)

These authors extend Owen's method. Hjort and Mohn (1984) used a 'cross' neighbourhood with locally straight boundaries and patterns obtained with probabilities

$$p \qquad q/4 \qquad r/4$$

so $p+q+r=1$. 'Don't know' classifications were allowed. The parameters p, q, r and any parameters in the marginal probabilities $\{\mathrm{P}(L_{ij}=k)\}$ and the noise are estimated from the data with the help of a training set for which the ground truth is known. An autocorrelated model for the noise ε was also considered, namely

$$\mathrm{cov}(\delta_{ij}, \varepsilon_{rs})=[\gamma\rho^{d(ij,rs)}+(1-\gamma)\delta_{ij=rs}]\Sigma$$

for $0<\gamma\leqslant 1$ and $0<\rho<1$. This is sufficient to compute $\mathrm{P}(L_{ij}=k\,|\,\mathbf{Z}_N)$. Saebø *et al.* (1985, p. 18) suggest $\rho\approx 0.94$, $\gamma\approx 0.81$ for some SPOT satellite data.

Hjort (1985a) extended this procedure to a 'box' 3×3 neighbourhood consisting of the central pixel and eight neighbours. The allowed patterns are now

$$\begin{array}{ccc} A & A & A \\ A & A & A \\ A & A & A \end{array} \qquad \begin{array}{ccc} A & A & B \\ A & A & A \\ A & A & A \end{array} \qquad \begin{array}{ccc} A & A & B \\ A & A & B \\ A & A & A \end{array}$$

$$\begin{array}{ccc} A & A & B \\ A & A & B \\ A & A & B \end{array} \qquad \begin{array}{ccc} A & B & B \\ A & A & B \\ A & A & A \end{array} \qquad \begin{array}{ccc} A & B & B \\ A & A & B \\ A & A & B \end{array}$$

and rotations and reflections of these, which gives rise to a six-parameter model for the prior. Again these parameters can be related to a single parameter for the Poisson mosaic model.

Hjort (1985a, b) points out that the parameters in the model of $\mathbf{l}_N$ can be estimated without training data. If a sparse sample of neighbourhoods is chosen, the model implies a parametric model for $\mathbf{Z}_N$ which can be fitted by maximum likelihood from the data for the samples (which are assumed to be independent).

These methods are tested on simulated remotely sensed data in Saebø *et al.* (1985), Hjort, Mohn and Storvik (1986) and Mohn, Hjort and Storvik (1986).

5.5.4 Haslett (1985)

Pickard (1977, 1980) defined a rather special binary MRF with the property that

$$P(l_{ij}=k \mid \text{rest of row and column containing } (i, j))$$

$$P(l_{ij}=k \mid l_N, l_E, l_S, l_W)$$

Haslett used this as a partial justification for restricting attention to a 'cross' of pixels and maximizing $P(L_{ij}=k|\mathbf{Z}_N)$. Further, for this MRF, l_N, l_E, l_S and l_W are conditionally independent given l_C. This enables (11) to be used directly with parameters the transition probabilities $P(l_N=j|l_C=i)$. Welch and Salter (1971) reached the same method from another point of view.

5.5.5 Markov mesh random fields (MMRF)

Markov mesh random fields (Abend, Harley and Kanal, 1965; Kanal, 1980) are a 'unilateral' subclass of Markov random fields. The pixels of our rectangular $M \times N$ array are partially ordered. Define

$$A_{i,j}=\{(k, l) \mid k<i \text{ or } l<j\}$$

$$B_{i,j}=\{(k, l) \mid k\leqslant i \text{ and } l\leqslant j\}\backslash\{(i, j)\}$$

Then a MMRF is defined by

$$P(l_{ij} \mid l \text{ in } A_{i,j})=P(l_{ij} \mid l \text{ in } C_{i,j} \subset B_{ij})$$

where $C_{i,j}$ determines the neighbour system used. This can be seen graphically

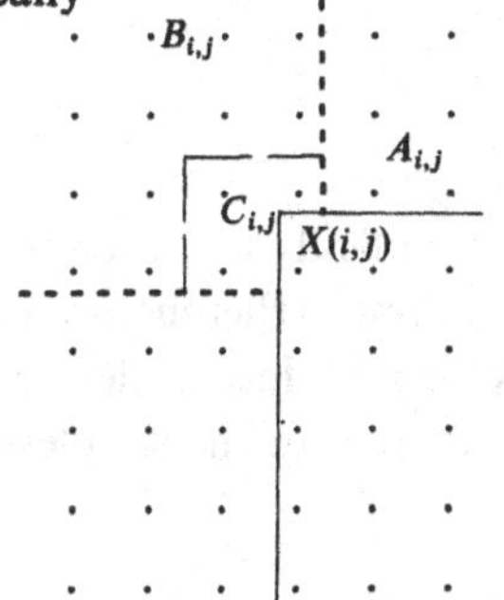

Examples of $C_{i,j}$ are $\{(i-1,j), (i,j-1)\}$ (sometimes called a second-order MMRF) and $\{(i-1,j), (i,j-1), (i-1,j-1)\}$, in both cases with suitable adjustments in the top row and left column. The advantages of MMRFs are that they are easy to simulate directly, and

$$\mathrm{P}(\mathbf{L}=\mathbf{l}) = \prod_{(i,j)} \mathrm{P}(L_{ij}=l_{ij} \mid L_a, a \in C_{i,j})$$

From this we see that MMRFs *are* Markov random fields and the two choices above correspond to the 4-neighbour and 8-neighbour systems respectively. However, MMRFs are not coextensive with MRFs and tend to be much more directional. They are not generally regarded as adequate models for spatial phenomena.

Derin *et al.* (1984) proposed the maximization of $\mathrm{P}(L_{ij}=k \mid \mathbf{Z})$ by regarding the columns of $\mathbf{L}$ as a vector Markov chain. This is true for a general MRF but computationally prohibitive. Instead, they use parts of the column, and find l_{ij} from calculations on a strip containing rows $i-1$, i and $i+1$. For a MMRF (but not a general MRF) the triples $V_j = (l_{i-1,j}; l_{i,j}; l_{i+1,j})$ form a Markov chain viewed with additive noise. This one-dimensional problem is quite well understood, and we can find the MAP estimators of the V_j from the corresponding observations in two passes. The estimates of $l_{i,\cdot}$ are retained and the strip process repeated for each row.

Devijver (1985) used a second-order MMRF and a recursive procedure to find the MAP estimator of l_{ij} given

$$\{Z_{(k,l)} \mid k \leqslant 1 \text{ and } l \leqslant j\}$$

or

$$\{Z_{(k,l)} \mid k \leqslant i+1,\ l \leqslant j+1 \text{ but } (k,l) \neq (i+1, j+1)\}$$

or

$$\{Z_{(k,l)} \mid k \leqslant i \text{ and } l \leqslant j \text{ or } k < i \text{ and } l = j+1\}$$

The last will be appropriate for real-time operation on a raster-scanned image. More information is available once the whole image is received. Devijver's experiments show the very directional nature of his prior assumptions for $\mathbf{l}$. This is perhaps acceptable in contexts where speed is more important than high fidelity.

5.5.6 Smoothing methods

All the methods considered so far explicitly allow for boundaries to pass near the pixel under consideration. Some earlier methods assumed that the boundaries were rare, so one would lose little by pooling information between a pixel and its neighbours. In the simplest idea we

Figure 5.19 Smoothing a non-spatial segmentation. (a) Signal is average of $\mathbf{Z}$ and $\mathbf{Z}^e$ with four neighbours. (b) Signal is $\frac{1}{4}\mathbf{Z}+\frac{3}{4}\mathbf{Z}^e$ with eight neighbours.

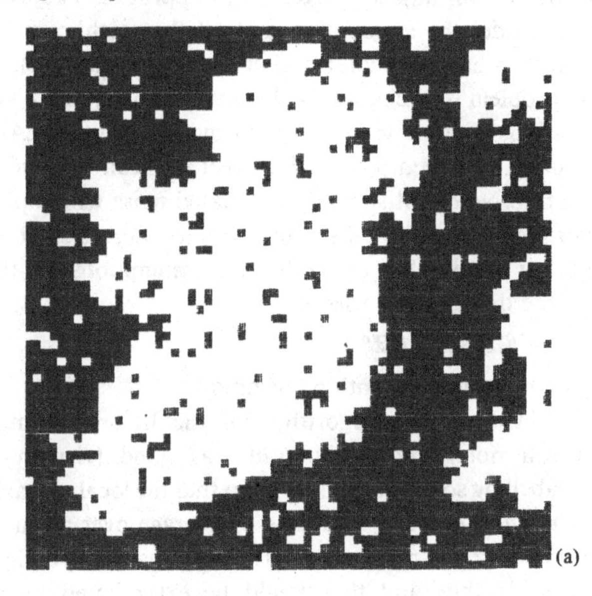

(a)

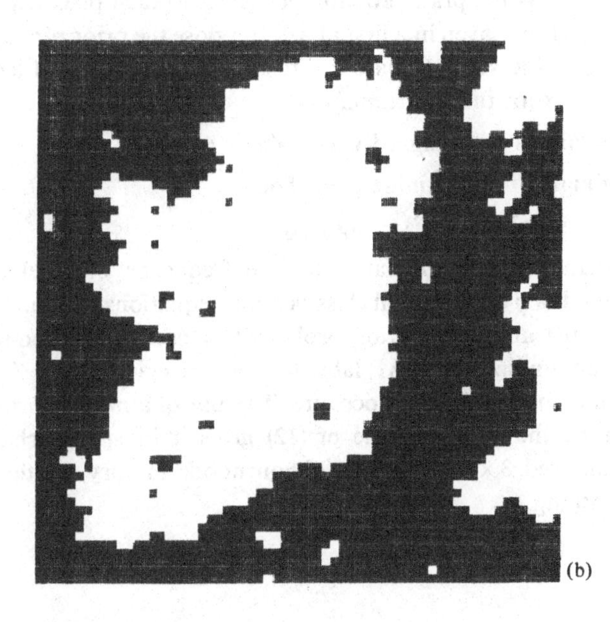

(b)

average Z_{ij} with observations on the neighbours of (i, j). Switzer (1980) proposed classifying l_{ij} on the basis of Z_{ij} and Z^e_{ij}, the average of the observations on the four adjacent pixels (with appropriate adjustment at the boundary). Under the assumption that all the neighbours have the same label one then has more information to choose that label.

The formal problem is to choose label k given data (Z_{ij}, Z^e_{ij}). We allow vector observations (Switzer was motivated by four-channel LANDSAT data), so assume each component has vector mean μ_k and known covariance matrix. Note that spatially correlated noise will imply that $\mathbf{Z}$ and $\mathbf{Z}^e$ are correlated. Standard discrimination techniques can be used to choose k. Switzer shows that under further assumptions on the noise structure that the discriminant used is

$$\mathbf{Z}^* = (\alpha - \beta)\mathbf{Z} + (1 - \beta)\mathbf{Z}^e$$

which is explicitly a signal-smoothing technique.

Figure 5.19 illustrates this algorithm on the Ireland example. The pictures are still noisy and nothing like as good (visually) as the simultaneous labelling solutions. It does seem that the local averaging may not be enough. An alternative would be to average over a much larger neighbourhood. Switzer's smaller examples show that smoothing distorts boundaries considerably, and this would be exacerbated by a larger neighbourhood.

Switzer (1983) referred to his 1980 technique as 'pre-smoothing'. An alternative, 'post-smoothing', is given by Switzer, Kowalik and Lyon (1982). This adjusts the prior probabilities given to each pixel depending on the classifications given in a first phase. Suppose the prior probabilities for pixel (i, j) in the second phase are to be $\pi_1(ij) \dots \pi_c(ij)$, and let f_{rs} be confusion matrix for the algorithm; that is

$$f_{rs} = P(\text{label } s \text{ is assigned when label } r \text{ is true})$$

which is estimated from training data. Then

$$P(\text{assign label } s \text{ to } ij) = \Sigma \pi_r(ij) f_{rs} \qquad (12)$$

The left-hand side is estimated by the frequency of label s in a neighbourhood of ij on the initial classification. Equations (12) can then be solved for $\pi_{.}(ij)$ and these prior probabilities used in a (non-spatial) discrimination, finding the MAP label for each pixel (i, j) given Z_{ij}.

This is an empirical Bayes procedure. The use of a neighbourhood of (i, j) estimating the left-hand side of (12) gives it its spatial character. Switzer *et al.* used 3×3 to 9×9 neighbourhoods to vary the degree of spatial smoothing.

6

Summarizing binary images

The type of images which occur in electron microscopy are rather different from those of remote sensing studied in chapter 5. First, much higher resolution is available, so 1024×1024 pixel pictures are not uncommon. In general resolution is limited by display considerations rather than the physical imaging process. Conversely, the information at each pixel is much more limited. Although greylevels may in theory be available, pictures are usually thresholded to be black or white. In some cases a limited number of colours are used. For example, my interest in these problems started with electron micrographs of cross-sections of rocks, in which the material of individual grains could be distinguished. There were six colours corresponding to the various component materials. (This is however not the norm.)

Fabbri (1984) provides an applied introduction to applications in geology. Commercial image analysers (such as the Quantimet) have found

Figure 6.1 Left and right halves of a 20 m × 10 m rectangular plot of ling (*Calluna vulgaris*), redigitized from Diggle (1981).

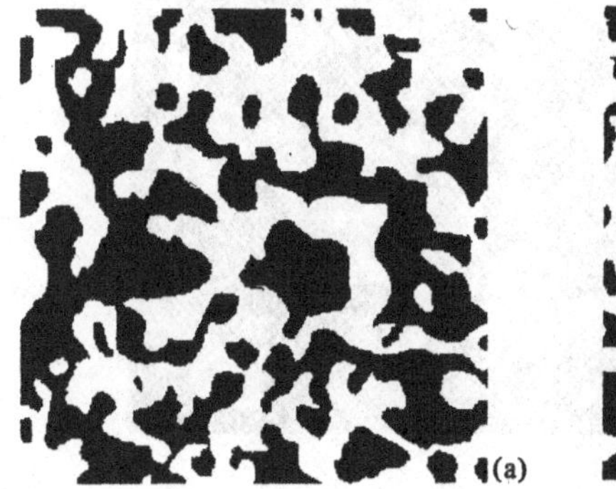

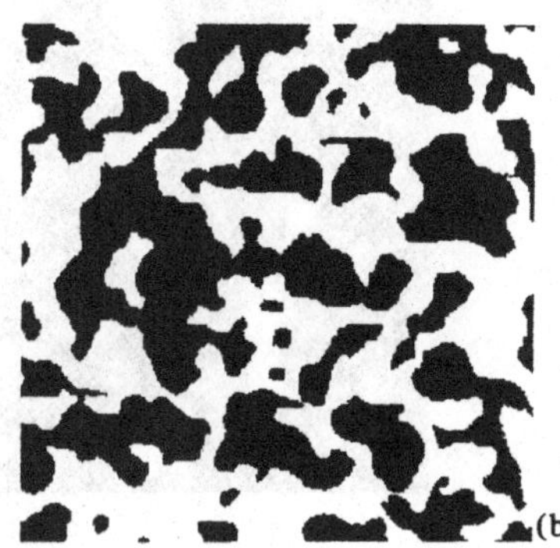

a wide range of uses including materials science and cell biology. Binary images can also arise from photographs and direct mapping of presence/absence. Figure 6.1 shows an image of presence or absence of ling (*Calluna vulgaris*) redigitized from Diggle (1981). The original was recorded on a 20 m × 10 m rectangle, here divided into 10 m × 10 m squares. Like all the examples shown in this chapter, the resolution is 256 pixels square. This is mainly for convenience, since cheap digitizers are available to grab video images at this resolution, and two images could be compared on the 720 × 560 pixel screen of the microcomputer used in this work. On more modern workstations one could routinely use 512 × 512 images. Because of the cheapness of the technology, applications of these techniques can be expected to mushroom.

Figures 6.2 and 6.3 show some further examples from geology and materials science respectively. All the examples are analysed below, after the theory has been developed. Much of this chapter is based on work published in Ripley (1986a, b).

Figure 6.2 An electron micrograph cross-section of smackover carbonate rock.

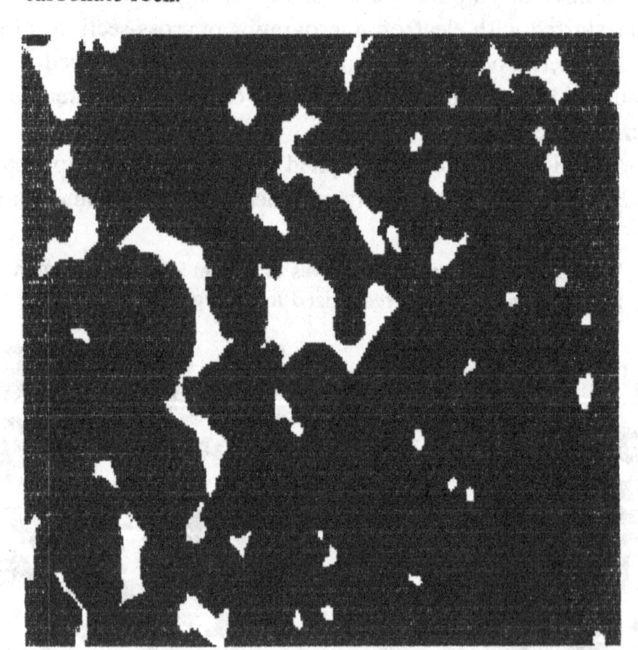

Figure 6.3 Cross-sections of glass fibre reinforced polyester (GFRP). The black represents glass, white the resin. (a) 2560 fibres at low magnification. (b) 50 fibres at higher magnification.

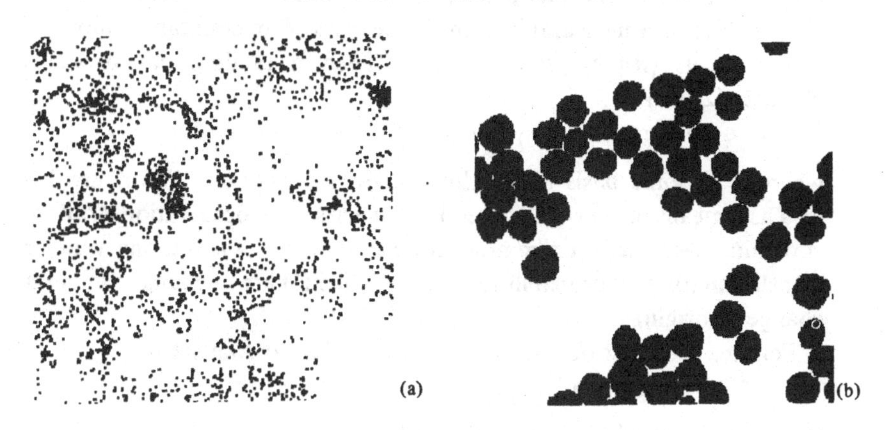

6.1 MATHEMATICAL FORMULATION

Since the underlying phenomena which we observe in a digital image are essentially continuous, our mathematical model should be based on $\mathbb{R}^2$. Let A denote the subset of $\mathbb{R}^2$ which gives rise to black pixels, so A^c gives rise to white pixels. Since A is a random subset of $\mathbb{R}^2$, we need a theory of random subsets of the plane. This has mathematical subtleties. A theory for general subsets has been proposed (Matheron, 1967) but many interesting observations are then not measurable (in the technical sense of measure theory). This led Matheron to restrict attention to random *closed* sets (Matheron, 1975). Parallel considerations led Kendall (1974) to a more abstract theory which specializes in our problem to a theory equivalent to Matheron's. Let us give some indication of what can go wrong. We could model figure 6.3a as either a point process or a random set, and the realization of a point process is a random set. We would like $N(E)$, the count of the number of points in E, to be a random variable. In Matheron's 1967 theory it is not, but in the closed set theories it is. (This is not quite trivial; it is proved in Ripley, 1976a.)

In fact all our digital observations will correspond to observing A at a finite number of points and so be measurable under either theory. We could therefore build our work on either model. It is more theoretically satisfying to use the random closed set model. This is based on the measurability of the events

$$\{A \uparrow T_i, \quad i=1,\ldots,m, \quad A \cap T_0 = \varnothing\} \tag{1}$$

where $\uparrow$ denotes 'hits', having a non-empty intersection. Here T_i belong to a class of 'test sets'. Matheron and Kendall started from different classes of test sets (compact sets and open balls), but in both theories the class of T's for which (1) is a measurable event is the class $\mathscr{T}$ of countable unions of compact sets (Ripley, 1981, p. 195). Then a random closed set is characterized by

$$Q(T)=P(A\cap T=\varnothing) \quad \forall T\in\mathscr{T}_0$$

where $\mathscr{T}_0$ is some 'basis' for $\mathscr{T}$, closed under finite unions. We will use this idea as a means of generating useful summaries of an observation $A\cap W$ of A within a window W. From now on we assume that A is stationary under translation (that is, probabilities of events like (1) are unchanged under a change of origin).

Perhaps the most obvious characteristic of a random set is

$$p=\mathrm{P}(\mathbf{x}\in A)=1-Q(\{\mathbf{x}\})$$

for any point $\mathbf{x}$. This can be estimated by

$$\hat{p}=\frac{\text{area}(A\cap W)}{\text{area}(W)}$$

where the area is computed by counting dots. For example, in figure 6.1 we have $\hat{p}=0.477$ and 0.479 for the left and right plots.

After this we can consider two dots. We find

$$\mathrm{P}(\text{dots } \mathbf{x} \text{ and } \mathbf{y} \text{ are both in } A)=c(\mathbf{y}-\mathbf{x})$$

where $c(\mathbf{h})=c(-\mathbf{h})$ by symmetry. Thus c is a kind of uncentred covariance. We can estimate it in an obvious way by the proportion of pairs of dots $\mathbf{x}$, $\mathbf{x}+\mathbf{h}$ which are both black. That is,

$$\hat{c}(\mathbf{h})=\frac{\text{area}\{\mathbf{x}\,|\,\mathbf{x}\in A\cap W\ \&\ \mathbf{x}+\mathbf{h}\in A\cap W\}}{\text{area}\{\mathbf{x}\,|\,\mathbf{x}\in W\ \&\ \mathbf{x}+\mathbf{h}\in W\}}$$

$$=\frac{\text{area}[(A\cap W)\cap\{(A\cap W)-\mathbf{h}\}]}{\text{area}[W\cap(W-\mathbf{h})]}$$

Note the edge correction used in the denominator, which can be described both as a border method and as a translation correction. Figure 6.4 shows $\hat{c}$ for horizontal and vertical directions on figure 6.1.

Unfortunately, $\hat{c}$ has little discriminatory power. Figure 6.5 shows a simulation of a model proposed by Diggle (1981) for the ling data together with its covariance summary. Whereas figure 6.5a appears quite different to figure 6.1a,b to most observers, the $\hat{c}$ summaries are indistinguishable. This failure of second-order methods in images has been observed elsewhere (Enting and Welberry, 1978).

What is needed is a measure of the *shape* of the observed image $A\cap W$.

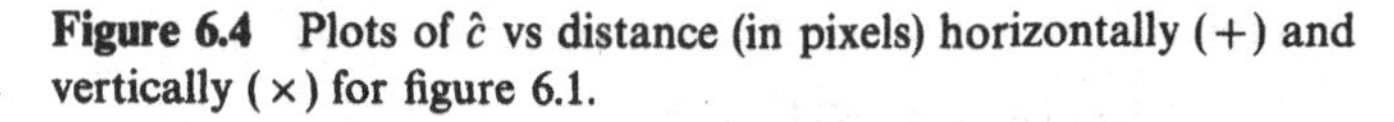

Figure 6.4 Plots of $\hat{c}$ vs distance (in pixels) horizontally (+) and vertically (×) for figure 6.1.

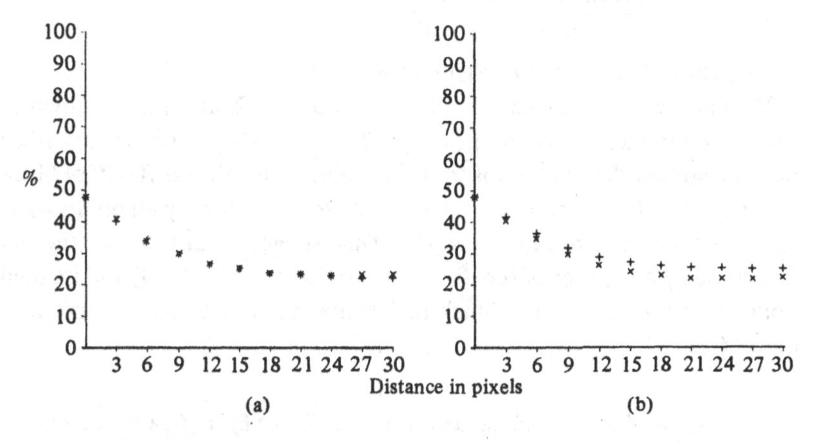

Figure 6.5 (a) Simulation of Diggle's model and (b) its $\hat{c}$ summary.

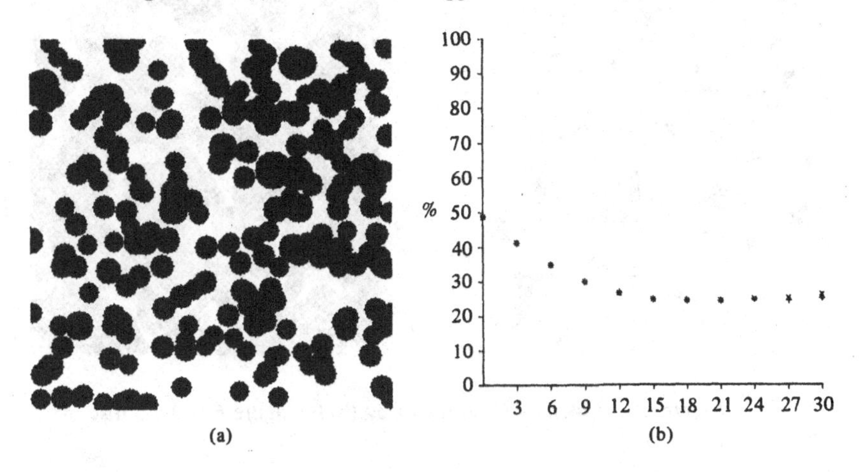

From the theory of random sets we consider $P(T \subset A^c)$ and $P(T \subset A)$. The former is $Q(T)$, but in the general the latter event is not even measurable. However, our digital test sets will be finite, and so $P(T \subset A)$ is defined. A crude but unbiased estimator of $P(T \subset A)$ is $1(T \subset A)$. (Here T is fixed, A is random.) By stationarity we can also use $1(T + \mathbf{x} \subset A)$ for any point $\mathbf{x}$. Averaging these unbiased estimators over all $\mathbf{x}$ for which they are observable, we obtain

$$e_T = \frac{\text{area}\{\mathbf{x} \mid T + \mathbf{x} \subset A \cap W\}}{\text{area}\{\mathbf{x} \mid T + \mathbf{x} \subset W\}} \tag{2}$$

as an unbiased estimator of $\mathrm{P}(T \subset A)$. In a dual way

$$d_T = \frac{\text{area}\{\mathbf{x} \mid T+\mathbf{x} \subset W, T+\mathbf{x} \uparrow A\}}{\text{area}\{\mathbf{x} \mid T+\mathbf{x} \subset W\}} \tag{3}$$

is an unbiased estimator of $\mathrm{P}(A \uparrow T) = 1 - Q(T)$.

We can interpret the sets in the numerator of (2) and (3) as shown in figure 6.6 for a disc T. We interpret $\{\mathbf{x} \mid T+\mathbf{x} \subset A \cap W\}$ as the set described by the centre of the disc as it is 'rolled around' inside the set $A \cap W$ of black points, and $\{\mathbf{x} \mid T+\mathbf{x} \subset W, T+\mathbf{x} \uparrow A\}$ as the set described by all points as its centre is moved around in $A \cap W$. This suggests considering the set described by any point of the disc as it is rolled around $A \cap W$, and its dual obtained by interchanging black and white. These sets are illustrated in figure 6.7.

Figure 6.6 Erosion (a) and dilatation (b) for figure 6.1a by a disc of radius 5 pixels.

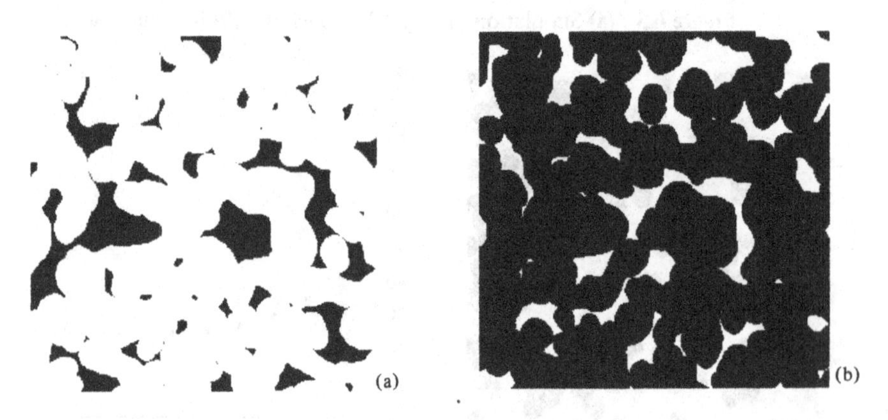

Figure 6.7 Opening (a) and closing (b) for figure 6.1a by a disc of radius 5 pixels.

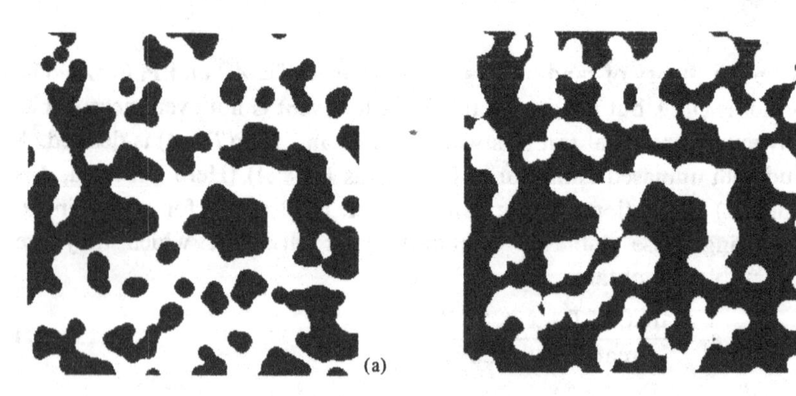

We can describe these sets more concisely using Serra's calculus (described in Matheron, 1975, and Serra, 1982). Define, for sets A and B,

$$\check{A}=\{-\mathbf{x}\,|\,\mathbf{x}\in A\}$$

$$A\oplus B=\{\mathbf{x}+\mathbf{y}\,|\,\mathbf{x}\in A,\ \mathbf{y}\in B\}$$

$$A\ominus B=(A^c\oplus B)^c=\{\mathbf{x}\,|\,B+\mathbf{x}\subset A\}$$

Then

$$e_T=\frac{\text{area}\{(A\cap W)\ominus\check{T}\}}{\text{area}\{W\ominus\check{T}\}}$$

and

$$d_T=\frac{\text{area}\{(A\oplus\check{T})\cap(W\ominus\check{T})\}}{\text{area}\{W\ominus\check{T}\}}$$

where the sets in the numerator were termed the 'erosion' and 'dilatation' of A by T by Serra, hence the notation. The sets in figure 6.7 are

$$(A\ominus\check{T})\oplus T$$

Serra's 'opening' of A by T, and

$$(A\oplus\check{T})\ominus T$$

the 'closing' of A by T. These operations are referred to as morphological transformations. We have to compute these from the observed part of A, $A\cap W$. Let V denote the maximal set such that the opening of A by T is known within V from $A\cap W$. By duality (interchanging the roles of black and white) this will be the appropriate set for closing as well.

Lemma

$$V=W\ominus(T\oplus\check{T})$$

Proof. From $A\cap W$ we know $A\ominus\check{T}$ and $A\oplus\check{T}$ on $W\ominus\check{T}$, as argued above. From

$$(A\ominus\check{T})\cap(W\ominus\check{T})=(A\cap W)\ominus\check{T}$$

we know $(A\ominus\check{T})\oplus T$ on

$$\begin{aligned}(W\ominus\check{T})\ominus T&=[(W\ominus\check{T})^c\oplus T]^c\\&=[(W^c\oplus\check{T})\oplus T]^c=[W^c\oplus(T\oplus\check{T})]^c\\&=W\ominus(T\oplus\check{T})\end{aligned}$$ □

This leads to two further quantities

$$o_T=\frac{\text{area}[\{(A\cap W)\ominus\check{T}\}\oplus T]\cap V}{\text{area } V}$$

$$c_T=\frac{\text{area}[\{(A\cap W)\oplus\check{T}\}\ominus T]\cap V}{\text{area } V}$$

6.2 A PROPOSED SUMMARY

Our proposed summary is to plot e_T, o_T, c_T and d_T against characteristics of T. Most of the examples are against the radius of a disc T. Figure 6.8 shows these summaries for the ling data and the simulation. The curve o_T clearly discriminates the data image from the simulation. We have

$$(A \ominus \check{T}) \subset (A \ominus \check{T}) \oplus T \subset (A \oplus \check{T}) \ominus T \subset A \oplus \check{T}$$

so

$$e_T \leqslant d_T, \qquad o_T \leqslant c_T$$

and

$$e_T \leqslant o_T \leqslant c_T \leqslant d_T$$

except perhaps for edge effects. Further, all four curves are monotone in T except for edge effects (e.g. o in figure 6.9). For discs T we can give a simpler explanation of the outer curves. Let $T = \mathrm{b}(\mathbf{0}, \lambda)$. Then

$$\begin{aligned} \mathrm{E}d_T &= \mathrm{P}(A \uparrow T) = \mathrm{P}(\exists\, \mathbf{x} \in A \text{ with } d(\mathbf{x}, \mathbf{0}) < \lambda) \\ &= \mathrm{P}(d(\mathbf{0}, A) < \lambda) \end{aligned}$$

for the Hausdorff distance from a point to a set. By stationarity we can replace $\mathbf{0}$ by any other point. Thus the d curve estimates the c.d.f. of the distance from a fixed point to the nearest point of A. Of course, this distribution has mass p at the origin. (In fact d is the c.d.f. of a random point in W to $A \cap W$, modulo edge effects.)

This explanation should be familiar from chapter 3, where for a point process we had

$$p(t) = \mathrm{P}(d(\mathbf{0}, A) < t)$$

where A is the set of points. Then $\hat{p}(t)$ as defined there is d_T for $T = b(\mathbf{0}, t)$.

By duality e_T is one minus the c.d.f. of the distance from a fixed point to A^c. No such simple interpretation of the middle curves is known. It is clear that

$$o_T \text{ estimates } \mathrm{P}[\mathbf{x} \in (A \text{ opened by } T)]$$

$$c_T \text{ estimates } \mathrm{P}[\mathbf{x} \in (A \text{ closed by } T)]$$

and that both sets appearing can be regarded as T-smoothed versions of A. The 'granulométrie bidimensionelle' defined by

$$F_A(t) = \begin{cases} o_{b(\mathbf{0}, |r|)} & \text{for } r \leqslant 0 \\ c_{b(\mathbf{0}, r)} & \text{for } r \geqslant 0 \end{cases}$$

and the 'répartition granulométrique' defined by

$$G_A(t) = \frac{o_\varnothing - o_{b(\mathbf{0}, t)}}{o_\varnothing}$$

Figure 6.8 Summaries for figure 6.1a (a), 6.1b (b) and 6.5a (c). In all seed plots the curves are d_T, c_T, o_T, e_T from top to bottom.

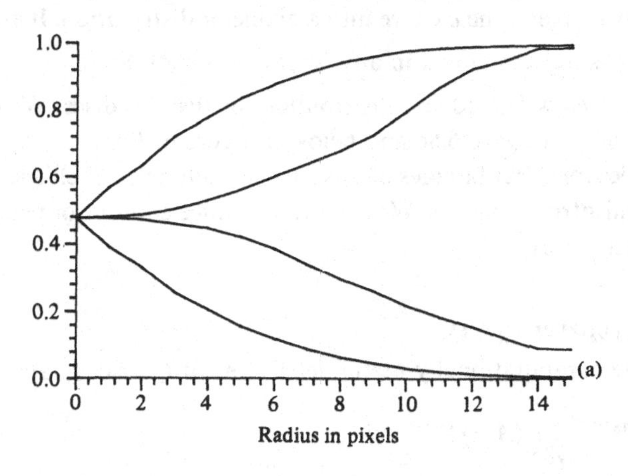

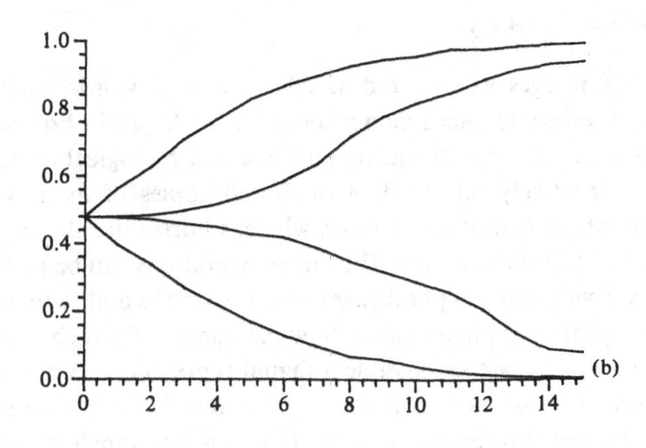

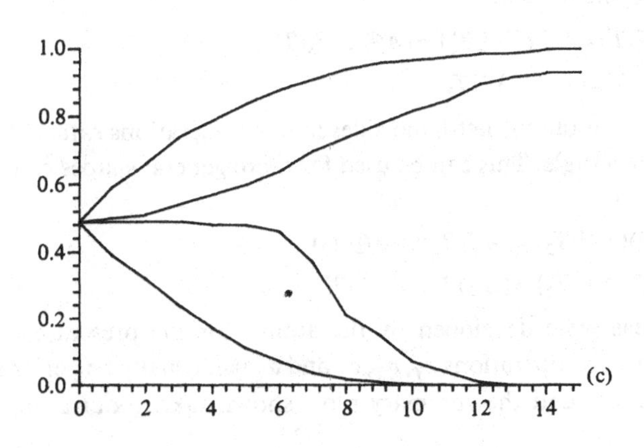

have been used in metallurgy (Fabbri, 1984, figure 4.3 is an example). If A is made up of non-overlapping discs of different diameters, it is clear that G_A correctly converts the o curve into a diameter distribution. It estimates

$$P(\mathbf{x} \text{ is not contained in any } b(\mathbf{y}, t) \cap A \mid \mathbf{x} \in A)$$

which is an area-weighted size distribution for disjoint discs. We can see this aspect of it in figure 6.8c and below in figure 6.10b.

Summaries for other families of sets such as rectangles or ellipses can be useful for anisotropic images. We can then produce contour or perspective plots as summaries.

6.3 COMPUTATION

The computationally useful definitions of $\ominus$ and $\oplus$ are

$$A \oplus T = \bigcup_{\mathbf{y} \in T} (A + \mathbf{y}) \tag{4}$$

$$A \ominus T = \bigcap_{\mathbf{y} \in T} (A + \mathbf{y}) \tag{5}$$

Our 256×256 images were stored as 256 rows of 8 words, each word containing 32 zeros and ones representing a line of 32 pixels. Formulae (4) and (5) show that $\oplus$ and $\ominus$ can be implemented by logical *or* and *and* operations respectively on $A \cap W$ with shifted copies of itself. Vertical shifting amounts to operating on rows, whereas horizontal shifting needs shifting the words a bit at a time. Thus these operations can be performed for arbitrary T with built-in parallelism from the CPU's ability to perform 32 bit-wise logical operations. This is how the images of this chapter were computed, using as exact as possible a digital representation of a disc.

For *rectangles* T we can do better. Let T_1 and T_2 be horizontal and vertical line segments respectively. Then $T_1 \oplus T_2$ is a rectangle T based on these line segments, and

$$A \oplus T = A \oplus (T_1 \oplus T_2) = (A \oplus T_1) \oplus T_2$$

$$A \ominus T = (A \ominus T_1) \ominus T_2$$

allow us to compute the left-hand sides in $w + h$ operations rather than wh, for a $w \times h$ rectangle. This can be used for more general shapes containing rectangles, for

$$A \oplus (T_1 \cup T_2) = (A \oplus T_1) \cup (A \oplus T_2)$$

$$A \ominus (T_1 \cup T_2) = (A \ominus T_1) \cap (A \ominus T_2) \quad .$$

These ideas were developed by the author for the present examples. With their use the operations e_T, o_T, c_T and d_T can usually be performed in about a second, and the summary plots shown take about a minute to calculate. Essentially the same ideas have been developed independently by Fabbri (1984, Appendix C), Pecht (1986) and Zhuang and Haralick (1986). An example of constructing a disc T from smaller sets is given by Sternberg (1984, figure 4).

6.4 EXAMPLES

For the ling data, we see in figure 6.8 that the summaries of the left and right halves of the experimental plot are remarkably similar, and so do appear to be expressing genuine shape characteristics of the pattern. The curves for figure 6.5 are rather different, with the exception of d_T which Diggle (1981) used in parameter estimation of his fitted model. That model is of a Poisson process of centres of discs with independent radii from a const + Weibull distribution. The disc-like nature shows up clearly in the o_T curve, even though the discs do overlap, and the more rapid decrease of e_T to zero shows that the white space is more evenly distributed. The c_T curve shows that the white phase is less 'smooth' for the simulation than for the data.

Many of these objections can be overcome by smoothing the simulation by a closing as shown in figure 6.9. Of course, p is now slightly too large, and ideally we would refit the parameters of the model to fit the smoothed realizations to the data. The summary based on figure 6.9a still shows some differences from the ling datasets: the simulation is still too regular.

The GFRP example in figures 6.3 should have the fibres reasonably uniformly spaced for strength, which is not the case in this specimen. For figure 6.3a which is essentially a point process, the top curve of figure 6.10a is an efficient way to compute $\hat{p}(\ )$. In the enlarged picture of 6.3b the curve o_T shows that the image is made up of discs of fairly constant radii. On the small figure (where all dots are two pixels by two pixels) the bottom two curves are essentially zero except for radii of one and two pixels. Presumably comparison of these summary plots for different specimens might reveal what characteristics do really determine strength of GFRP.

Figure 6.3 is a cross-section of smackover carbonate rock, which is known to be made up of the shells of ellipsoidal animals. This could be seen fairly clearly in the original micrograph (before thresholding to black or white), and explains the peculiar shapes of the pore boundaries. The summary in figure 6.12 is not too illuminating by itself. Figure 6.11 shows a simulated disc packing model of interpenetrating discs, chosen to match the summaries as closely as possible. The centres of the discs are a hard-core point process, with discs which are larger than the inhibition distance. The match of the summaries is quite good except for the e_T curve, which

Figure 6.9 Smoothed version of figure 6.5a (a) and its summary (b).

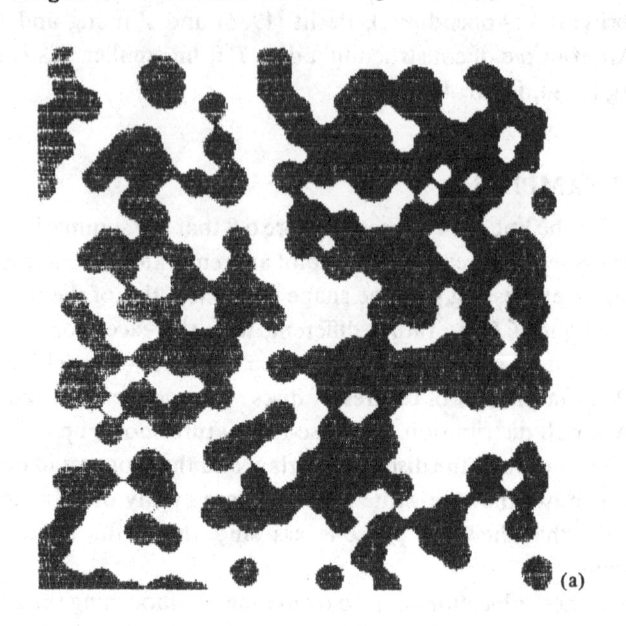

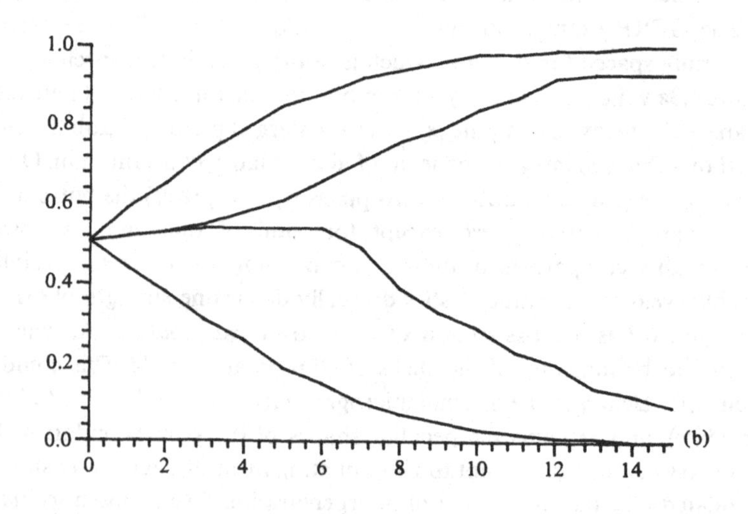

Figure 6.10 Summaries for the GFRP images of figure 6.3.

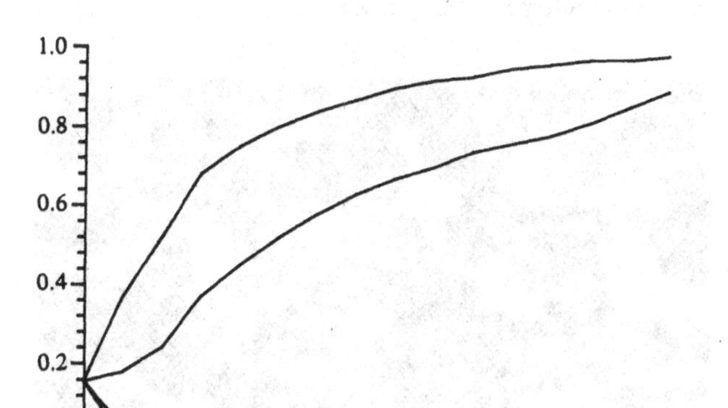

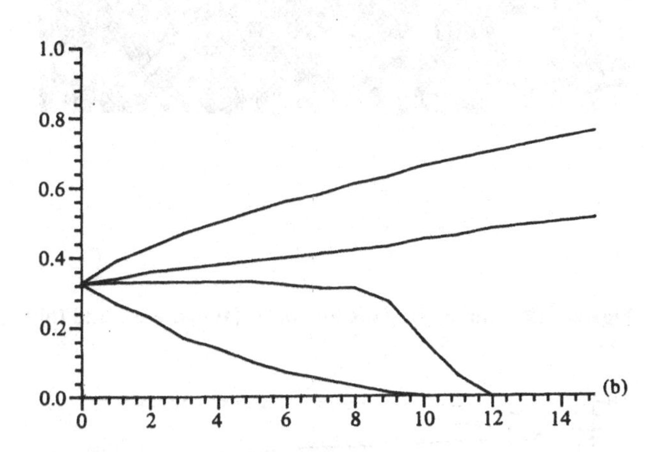

decreases more rapidly for the synthetic image. Remember then $1-e_T$ gives the c.d.f. of the distance from a fixed point to the white phase. Thus the real data exhibits heterogeneity compared to the synthetic one. This is in fact clear from figure 6.2, although the author did not notice this until after studying the summaries. In figure 6.2 the pores are much larger in the upper left part of the picture.

In both the ling data and the smackover rock we have the question of whether the observed summaries from the model and the data really are different. We can repeat the simulations of the models involved, and obtain

Figure 6.11 A simulated rock cross-section formed by random packing of interpenetrating discs.

Figure 6.12 Summary plots for rocks; (a) figure 6.1 and (b) figure 6.11.

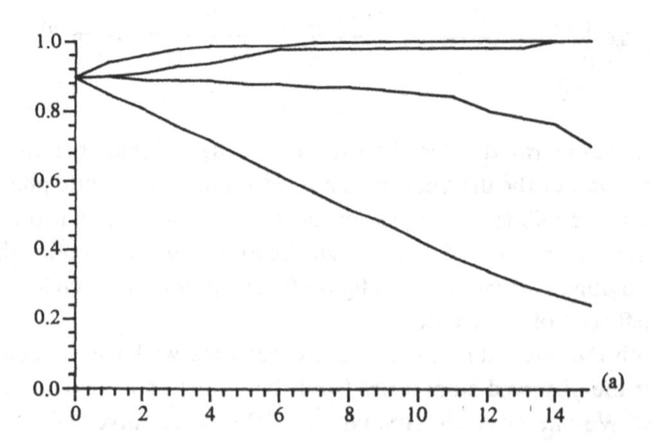

Figure 6.12 Continued

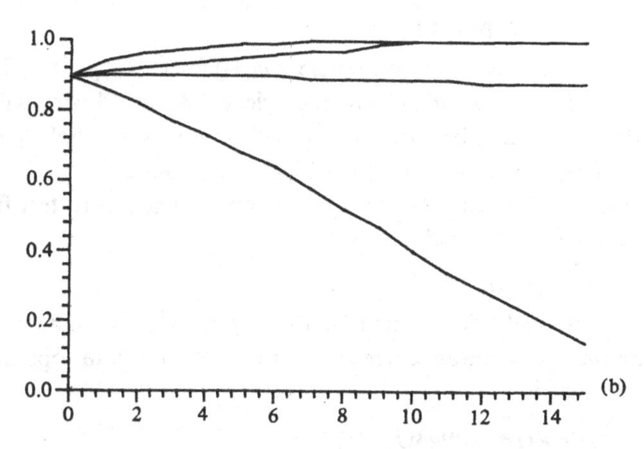

an envelope of the summary curves for the simulations, exactly as done for graphical summaries for point processes (Ripley, 1977). In both cases this confirms the lack of fit of the model.

6.5 EXTENSIONS

The random set theory on which these methods are based does not extend readily to non-binary images. For example, if we consider k-phase images there will be topological problems with what to do about the boundaries. Some progress is possible by comparing pairs of phases and assuming that the boundary is negligible. (The latter is a fair assumption in practice, but is theoretically intractable.) The impetus for further work in this area may come from technology as colour image analysis systems fall in price. Fabbri, Kasvand and Masounave (1983) and Fabbri (1984) consider patterns of contiguous grains of different types, a good example of a k-phase image. Their analysis is mainly of the contiguity between pairs of types of grains.

It is straightforward to extend the transformations used here to greylevel images. The set A is replaced by a function $f:\mathbb{R}^2 \to [0, \infty)$ giving the brightness at each point. Define

$$A_\lambda = \{\mathbf{x} \mid f(\mathbf{x}) \geqslant \lambda\} \tag{6}$$

Then A_λ corresponds to *thresholding* the image at brightness λ. (For technical reasons, the bright areas form A, represented by black above.) Then A_λ is a monotone family of sets, with $A_0 = \mathbb{R}^2$ and $A_\infty = \varnothing$. Now consider applying a morphological transform g (opening, closing, erosion, dilatation or a combination thereof) to f. We can apply g to every member of the family A_λ, and define $g(f)$ by

$$g(f)(\mathbf{x}) = \sup\{\lambda \mid \mathbf{x} \in g(A_\lambda)\} \tag{7}$$

Since the operations are monotonic, $\mathbf{x} \in g(A_\lambda)$ for all $\lambda < g(f)(\mathbf{x})$. This gives us notions of opening and closing of greylevel images. Closing will remove small dark patches whereas opening will remove small light patches, in both cases replacing them by their local background.

Sternberg (1984) extends this notion from test sets B to test functions. Define the *umbra* of $f: \mathbb{R}^d \to \mathbb{R}$ by

$$Uf = \{(\mathbf{x}, \lambda) \mid \lambda \leqslant f(\mathbf{x})\} \subset \mathbb{R}^{d+1}$$

Then f is uniquely determined by its umbra. In particular, unions and intersections of umbras correspond to max and min operations on functions, that is

$$Uf_1 \cup Uf_2 = U(\max(f_1, f_2))$$

$$Uf_1 \cap Uf_2 = U(\min(f_1, f_2))$$

Consider how $f \oplus B$ as defined above is related to its umbra. We see

$$(\mathbf{x}, \lambda) \in U(f \oplus B)$$
$$\text{iff } \mathbf{x} \in A_\lambda \oplus B$$
$$\text{iff } \mathbf{x} \in \bigcup\{\mathbf{y} \in B | \mathbf{x} - \mathbf{y} \in A_\lambda\}$$
$$\text{iff } \mathbf{x} \in \bigcup\{\mathbf{y} \in B | f(\mathbf{x} - \mathbf{y}) \geqslant \lambda\}$$

Thus

$$f \oplus B(\mathbf{x}) = \sup\{\mathbf{y} \in B \mid f(\mathbf{x} - \mathbf{y})\} \tag{8}$$

and

$$U(f \oplus B) = Uf \oplus UI_B \tag{9}$$

where I_B is the indicator function of B, defined here as zero for points in B, $-\infty$ for points in B^c.

We can replace I_B by an arbitrary function k. Then

$$(f \oplus k)(\mathbf{x}) = \sup\{\mathbf{y} \mid f(\mathbf{x} - \mathbf{y}) + k(\mathbf{y})\}$$

corresponds to the umbra of $Uf \oplus Uk$, generalizing (6–9). From duality, using (5), we find

$$(f \ominus k)(\mathbf{x}) = \inf\{\mathbf{y} \mid f(\mathbf{x} - \mathbf{y}) - k(\mathbf{y})\}$$

and from these we can form openings and closings by test functions k. Further details and applications are given by Sternberg (1986) and Haralick, Sternberg and Zhuang (1987).

Morphological transformations (both binary and greylevel) can be used within image analysis in a wide variety of ways. One can use specific transformations to screen images for objects of specific size and shape, and so build systems to inspect manufacturing lines for foreign bodies. There is a whole class of possible ways to use transformations in conjunction with the image analysis problems of chapter 5. They can be used within the definition of the prior, as part of the formal loss function, or as part of the assessment procedure for competing algorithms. After a long period of very specialized use, morphological transformations are beginning to take their rightful place as part of the 'exploratory data analysis' of images.

References

Abend, K., Harley, T. J. & Kanal, L. N. (1965). Classification of binary random patterns. *IEEE Trans.* **IT-11**, 538–44.

Baddeley, A. (1980). A limit theorem for some statistics of spatial data. *Adv. Appl. Probab.* **12**, 447–61.

Baddeley, A. J. & Silverman, B. W. (1984). A cautionary example on the use of second-order methods for analyzing point patterns. *Biometrics* **40**, 1089–93.

Besag, J. (1974). Spatial interaction and the statistical analysis of lattice systems. *J. Roy. Statist. Soc. B* **36**, 192–236.

Besag, J. (1975). Statistical analysis of non-lattice data. *The Statistician* **24**, 179–95.

Besag, J. (1977a). Some methods of statistical analysis for spatial data. *Bull. Int. Statist. Inst.* **47(2)**, 77–92.

Besag, J. (1977b). Errors-in-variables estimation for Gaussian lattice schemes. *J. Roy. Statist. Soc. B* **39**, 73–8.

Besag, J. (1981). On a system of two-dimensional recurrence equations. *J. Roy. Statist. Soc. B* **43**, 302–9.

Besag, J. (1983). Discussion of invited papers. *Bull. Int. Statist. Inst.* **50(3)**, 422–5.

Besag, J. (1986). On the statistical analysis of dirty pictures. *J. Roy. Statist. Soc. B* **48**, 259–302.

Besag, J. & Kempton, R. (1986). Statistical analysis of field experiments using neighbouring plots. *Biometrics* **42**, 231–51.

Besag, J., Milne, R. & Zachary, S. (1982). Point process limits of lattice processes. *J. Appl. Prob.* **19**, 210–16.

Brown, D. & Rothery, P. (1978). Randomness and local regularity of points in a plane. *Biometrika* **65**, 115–22.

Brown, T. C. & Silverman, B. W. (1979). Rates of Poisson convergence for *U*-statistics. *J. Appl. Probab.* **16**, 428–32.

Clark. P. J. & Evans, F. C. (1954). Distance to nearest neighbour as a measure of spatial relationships in populations. *Ecology* **35**, 445–53.

Cook, D. G. & Pocock, S. J. (1983). Multiple regression in geographical morbidity studies, with allowance for spatially autocorrelated errors. *Biometrics* **39**, 361–71.

Dahlhaus, R. & Künsch, H. (1987). Edge effects and efficient parameter estimation for stationary random fields. *Biometrika* **74**, 887–82.

Daley, D. J. & Vere-Jones, D. (1972). A summary of the theory of point processes. In *Stochastic Point Processes*, ed. P. A. W. Lewis, Wiley, New York, 299–383.

Derin, H., Elliott, H., Cristi, R. & Geman, D. (1984). Bayes smoothing algorithms for segmentation of binary images modeled by Markov random fields. *IEEE Trans.* **PAMI-6**, 707–20.

Devijver, P. A. (1985). Probabilistic labeling in a hidden second order Markov mesh. Phillips Research Lab. report.

Devijver, P. A. & Kittler, J. V. (1982). *Pattern Recognition: A Statistical Approach.* Prentice-Hall, Engelwood Cliffs, NJ.

De Vos, S. (1973). The use of nearest neighbour distances. *Tijdschrift voor Economische en Sociale Geografie* **64**, 307–19.

Diggle, P. J. (1979). On parameter estimation and goodness-of-fit testing for spatial point patterns *Biometrics* **35**, 87–101.

Diggle, P. J. (1981). Binary mosaics and the spatial pattern of heather. *Biometrics* **37**, 531–9.

Diggle, P. J. (1983). *Statistical Analysis of Spatial Point Patterns.* Academic Press, London.

Donnelly, K. P. (1978). Simulations to determine the variance and edge effect of total nearest neighbour distance. In *Simulation Methods in Archaeology*, ed. I. Hodder. Cambridge University Press, Cambridge.

Eberl, W. & Hafner, R. (1971). Die asymptotische Verteilung von Koinzidenzen. *Z. f. Wahr.* **18**, 322–32.

Enting, I. G. & Welberry, T. R. (1978). Connections between Ising models and various probability distributions. *Suppl. Adv. Appl. Probab.* **10**, 65–72.

Fabbri, A. G. (1984). *Image Processing of Geological Data.* Van Nostrand Reinhold, New York.

Fabbri, A. G., Kasvand, T. & Masounave, J. (1983). Adjacent relationships in aggregates of crystal profiles. In *Pictorial Data Analysis*, ed. R. M. Haralick, NATO ASI **F4**, Springer, Berlin, 449–68.

Fiksel, T. (1984). Estimation of parameterized pair potentials of marked and non-marked Gibbsian point processes. *Elektron. Inform. Kybernet.* **20**, 270–8.

Frieden, B. R. (1979). Image enhancement and reconstruction. In *Picture Processing and Digital Filtering*, ed. T. S. Huang, Springer, Berlin, 177–248.

Geman, S. & Geman, D. (1984). Stochastic relaxation, Gibbs distributions and the Bayesian restoration of images. *IEEE Trans.* **PAMI-6**, 721–41.

Gidas, B. (1985). Nonstationary Markov chains and convergence of the annealing algorithm. *J. Statist. Phys.* **39**, 73–131.

Glass, L. & Tobler, W. R. (1971). Uniform distribution of objects in a homogeneous field: Cities on a plain. *Nature* **233 (5314)**, 67–8.

Good, I. J. & Gaskins, R. A. (1971). Nonparametric roughness penalties for probability densities. *Biometrika* **58**, 255–77.

Grenander, U. (1983). *Tutorial in Pattern Theory*. Division of Applied Mathematics, Brown University.

Guild, F. J. & Silverman, B. W. (1978). The microstructure of glass fibre reinforced polyester resin composites. *J. Microscopy* **114**, 131–41.

Gull, S. F. & Skilling, J. (1985). The entropy of an image. In *Maximum-Entropy and Bayesian Methods in Inverse Problems*, eds. C. R. Smith & W. T. Gandy Jr., Reidel, Dordrecht, 287–301.

Guyon, X. (1982). Parameter estimation for a stationary process on a *d*-dimensional lattice. *Biometrika* **69**, 95–105.

Hanisch, K.-H. (1984). Some remarks on estimators of the distribution function of nearest neighbour distance in stationary spatial point processes. *Math. Oper. Statist. ser. Statist.* **15**, 409–12.

Haralick, R. M., Sternberg, S. R. & Zhuang, X. (1987). Image analysis using mathematical morphology. *IEEE Trans.* **PAMI-9**, 532–50.

Haslett, J. (1985). Maximum likelihood discrimination in the plane using a Markovian model of spatial context. *Patt. Recgn.* **18**, 287–96.

Hjort, N. L. (1985a). Neighbourhood based classification of remotely sensed data based on geometric probability models. In Saebø *et al.* (1985).

Hjort, N. L. (1985b). Estimating parameters in neighbourhood based classifiers for remotely sensed data, using unclassified vectors. In Saebø *et al*, (1985).

Hjort, N. L. & Mohn, E. (1984). A comparison of some contextual methods in remote sensing classification. *Proc. 18th Int. Symp. Remote Sens. Env., CNES, Paris.* (Also in Saebø *et al.*, 1985.)

Hjort, N. L., Mohn, E. & Storvik, G. (1986). A simulation study of some contextual classification methods for remotely sensed data. *IGARSS '86*, Zürich.

Hsu, S. Y. & Mason, J. D. (1974). The nearest neighbor statistics for testing randomness of point distributions in a bounded two-dimensional space. *Proc. 1972 Meeting IGU Comm. Quant. Geogr.*, ed. M. H. Yeates, McGill-Queens University Press, Montreal.

Jaynes, E. T. (1985). Entropy and search theory. In *Maximum-Entropy and Bayesian Methods in Inverse Problems*, eds. C. R. Smith & W. T. Gandy Jr., Reidel, Dordrecht, 443–54.

Kanal, L. N. (1980). Markov mesh models. *Comp. Graph. Image. Proc.* **12**, 371–5.

Kelly, F. P. & Ripley, B. D. (1976). On Strauss's model for clustering. *Biometrika* **63**, 357–60.

Kendall, D. G. (1974). Foundations of a theory of random sets. In *Stochastic Geometry*, eds. E. F. Harding & D. G. Kendall, Wiley, Chichester, 322–76.

Kester, A. (1975). Asymptotic normality of the number of small distances between random points in a cube. *Stoch. Proc. Appl.* **3**, 45–54.

Kindermann, R. & Snell, J. L. (1980). *Markov Random Fields and Their Applications.* Amer. Math. Soc., Providence, RI.

Kirkpatrick, S., Gellatt, C. D. & Vecchi, M. P. (1983). Optimization by simulated annealing. *Science* **220**, 671–80.

Kittler, J. & Foglein, J. (1984). Contextual classification of multispectral pixel data. *Image Vision Comput.* **2**, 13–29.

Künsch, H. R. (1987). Intrinsic autoregressions and related models on the lattice $\mathbb{Z}^2$. *Biometrika* **74**, 517–24.

Lotwick, H. W. & Silverman, B. W. (1982). Methods for analysing spatial processes of several types of points. *J. Roy. Statist. Soc. B* **44**, 406–13.

Mardia, K. V. & Marshall, R. J. (1984). Maximum likelihood estimation of models for residual covariance in spatial regression. *Biometrika* **71**, 135–46.

Marroquin, J., Mitter, S. & Poggio, T. (1987). Probabilistic solution of ill-posed problems in computational vision. *J. Amer. Statist. Assoc.* **82**, 76–89.

Matérn, B. (1972). Poisson processes in the plane and related models for clumping and heterogeneity. NATO ASI on Statistical Ecology, Penn. State. Univ.

Matheron, G. (1967). *Éléments pour un Théorie des Milieux Poreaux.* Masson, Paris.

Matheron, G. (1973). The intrinsic random functions and their applications. *Adv. Appl. Probab.* **5**, 439–68.

Matheron, G. (1975). *Random Sets and Integral Geometry.* Wiley, New York.

Metropolis, N., Rosenbluth, A. W., Rosenbluth, M. N., Teller, A. H. & Teller, E. (1953). Equations of state calculations by fast computing machines. *J. Chem. Phys.* **21**, 1087–92.

Miles, R. E. (1974). On the elimination of edge effects in planar sampling. In *Stochastic Geometry*, eds. E. F. Harding & D. G. Kendall, Wiley, Chichester, 228–47.

Mitra, D., Romeo, F. & Sangiovanni-Vincentelli, A. (1986). Convergence and finite-time behavior of simulated annealing. *Adv. Appl. Probab.* **3**, 747–71.

Mohn, E., Hjort, N. L. & Storvik, G. (1986). A comparison of some classification methods in remote sensing by a Monte Carlo study. Norwegian Computing Centre Note **kart/03/86**.

Ogata, Y. & Tanemura, M. (1981). Estimation of interaction potentials of spatial point patterns through the maximum likelihood procedure. *Ann. Inst. Statist. Math.* **33B**, 315–38.

Ogata, Y. & Tanemura, M. (1984). Likelihood analysis of spatial point patterns. *J. Roy. Statist. Soc. B.* **46**, 496–518.

Ohser, J. (1983). On estimators for the reduced second moment measure of point processes. *Math. Oper. Statist. ser. Statist.* **14**, 63–71.

Ohser, J. & Stoyan, D. (1981). On the second-order and orientation analysis of planar stationary point processes. *Biom. J.* **23**, 523–33.

Owen, A. (1984). A neighbourhood-based classifier for LANDSAT data. *Can. J. Statist.* **12**, 191–200.

Owen, A. (1986). Discussion of Ripley (1986b). *Can. J. Statist.* **14**, 106–10.

Pecht, J. (1986). Speeding-up successive Minkowski operations with bit-plane computers. *Patt. Recgn. Lett.* **4**, 113–17.

Penttinen, A. (1984). Modelling interaction in spatial point patterns: parameter estimation by the maximum likelihood method. *Jyväskylä Studies in Computer Science, Economics and Statistics*, **7**.

Persson, O. (1972). The border effect on the distance between sample point and closest individual in the square. *IUFRO Third Conference Advisory Group Forest Statisticians.* INRA, Paris, 241–6.

Pickard, D. K. (1977). A curious binary lattice process. *J. Appl. Probab.* **14**, 717–31.

Pickard, D. K. (1980). Unilateral Markov fields. *Adv. Appl. Prob.* **12**, 655–71.

Pickard, D. K. (1987). Inference for discrete Markov random fields: the simplest nontrivial case. *J. Amer. Statist. Assoc.* **82**, 90–6.

Pincus, M. (1968). A closed form solution of certain programming problems *Oper. Res.* **16**, 690–4.

Pincus, M. (1970). A Monte-Carlo method for the approximate solution of certain types of constrained optimization problems. *Oper. Res.* **18**, 1225–8.

Ripley, B. D. (1976a). Locally finite random sets: Foundations for point process theory. *Ann. Probab.* **4**, 983–94.

Ripley, B. D. (1976b). The second-order analysis of stationary point processes. *J. Appl. Probab.* **13**, 255–66.

Ripley, B. D. (1977). Modelling spatial patterns. *J. R. Statist. Soc. B* **39**, 172–212.

Ripley, B. D. (1979). Tests of 'randomness' for spatial point patterns. *J. Roy. Statist. Soc. B.* **41**, 368–74.

Ripley, B. D. (1981). *Spatial Statistics.* Wiley, New York.

Ripley, B. D. (1982). Edge effects in spatial stochastic processes. In *Statistics in Theory and Practice. Essays in Honour of Bertil Matérn*, ed. B. Ranneby, Swedish University of Agricultural Sciences, Umeå, 242–62.

Ripley, B. D. (1984a). Edge corrections for spatial processes. In *Stochastic Geometry, Geometrical Statistics, Stereology*, eds. R. V. Ambartzumian & W. Weil, Teubner Texte, Leipzig, 144–53.

Ripley, B. D. (1984b). Contribution to the discussion of Drs Diggle & Gratton's paper. *J. Roy. Statist. Soc. B* **46**, 222.

Ripley, B. D. (1985). Analyses of nest spacings. In *Statistics in Ornithology*, eds. B. J. T. Morgan & P. M. North. Springer Lecture Notes in Statistics, **28**, 151–8.

Ripley, B. D. (1986a). Summaries for digital images. In *Geobild 85*, ed. W. Nagel. FSU, Jean, 6–9.

Ripley, B. D. (1986b). Statistics, images and pattern recognition. *Can. J. Statist.* **14**, 83–111.

Ripley, B. D. (1987). *Stochastic Simulation.* Wiley, New York.

Ripley, B. D. & Kelly, F. P. (1977). Markov point processes. *J. Lond. Math. Soc.* **15**, 188–92.

Ripley, B. D. & Silverman, B. W. (1978). Quick tests for spatial interaction. *Biometrika* **65**, 641–2.

Ripley, B. D. & Taylor, C. C. (1987). Pattern recognition. *Science Progress* **71**, 413–28.

Saebø, H., Bråten, K., Hjort, N. L., Llewellyn, B. & Mohn, E. (1985). *Contextual Classification of Remotely Sensed Data: Statistical Methods and Development of a System.* Norwegian Computing Centre Report **768**.

Saunders, R. & Funk, G. M. (1977). Poisson limits for a clustering model of Strauss. *J. Appl. Probab.* **14**, 776–84.

Saunders, R., Kryscio, R. J. & Funk, G. M. (1982). Poisson limits for a hard-core clustering model. *Stoc. Proc. Applic.* **12**, 97–106.

Serra, J. (1982). *Image Analysis and Mathematical Morphology.* Academic Press, London.

Silverman, B. W. (1976). Limit theorems for dissociated random variables. *Adv. Appl. Probab.* **8**, 806–19.

Silverman, B. W. (1978). Distances on circles, toruses and spheres. *J. Appl. Probab.* **15**, 136–43.

Silverman, B. W. & Brown, T. C. (1978). Short distances, flat triangles and Poisson limits. *J. Appl. Probab.* **15**, 815–25.

Skilling, J. & Gull, S. F. (1985). Algorithms and Applications. In *Maximum-Entropy and Bayesian Methods in Inverse Problems*, eds. C. R. Smith & W. T. Gandy Jr., Reidel, Dordrecht, 83–132.

Sternberg, S. R. (1984). Esoteric iterative algorithms. In *Digital Image Analyses*, ed. S. Levialdi. Pitman, London, 60–8.

Sternberg, S. R. (1986). Grayscale morphology. *Comput. Vision Graph. Imag. Proc.* **35**, 333–55.

Strauss, D. J. (1975). A model for clustering. *Biometrika* **63**, 467–75.

Strauss, D. J. (1977). Clustering on coloured lattices. *J. Appl. Probab.* **14**, 135–43.

Swain, P. H., Vardeman, S. B. & Tilton, J. C. (1981). Contextual classification of multispectral data. *Patt. Recgn.* **13**, 429–41.

Switzer, P. (1965). A random set process in the plane with a Markov property. *Ann. Math. Statist.* **36**, 1859–63.

Switzer, P. (1980). Extensions of linear discriminant analysis for statistical classification of remotely sensed satellite imagery. *Math. Geol.* **12**, 367–76.

Switzer, P. (1983). Some spatial statistics for the interpretation of satellite data. *Bull. Int. Statist. Inst.* **50(2)**, 962–7.

Switzer, P., Kowalik, W. S. & Lyon, R. J. P. (1982). A prior probability method for smoothing discriminant classification maps. *Math. Geol.* **14**, 433–44.

Takacs, R. (1986). Estimator for the pair potential of a Gibbsian point process. *Statistics* **17**, 429–33.

Titterington, D. M. (1985). Common structure of smoothing techniques in statistics. *Int. Statist. Rev.* **53**, 141–70.

Vere-Jones, D. (1978). Space time correlations for microearthquakes–a pilot study. *Suppl. Adv. Appl. Probab.* **10**, 73–87.

Vincent, P. J., Sibley, D., Ebdon, D. & Charlton, B. (1976). Methodology by example: Caution towards nearest neighbours. *Area* **8**, 161–71.

Warnes, J. J. (1986). A sensitivity analysis for universal kriging. *Math, Geology* **18**, 653–76.

Warnes, J. J. & Ripley, B. D. (1987). Problems with likelihood estimation of covariance functions of spatial Gaussian processes. *Biometrika* **74**, 640–2.

Welch, J. R. & Salter, K. G. (1971). A context algorithm for pattern recognition and image interpretation. *IEEE Trans.* **SMC-1**, 24–30.

Wohlberg, G. & Pavlidis, T. (1985). Restoration of binary images using stochastic relaxation with annealing. *Patt. Recgn. Lett.* **3**, 375–88.

Whittle, P. (1954). On stationary processes in the plane. *Biometrika* **41**, 434–49.

Zhuang, X. & Haralick, R. M. (1986). Morphological structuring element decomposition. *Comput. Vision Graph. Imag. Proc.* **35**, 370–82.

Index

Printed in the United States
By Bookmasters